SECOND EDITION

APPLICATIONS OF FIRE RESEARCH AND IMPROVEMENT

Mic Gunderson

Jeffrey T. Lindsey

JONES & BARTLETT LEARNING

World Headquarters
Jones & Bartlett Learning
5 Wall Street
Burlington, MA 01803
978-443-5000
info@jblearning.com
www.jblearning.com

Jones & Bartlett Learning books and products are available through most bookstores and online booksellers. To contact Jones & Bartlett Learning Public Safety Group directly, call 800-832-0034, fax 978-443-8000, or visit our website,

Substantial discounts on bulk quantities of Jones & Bartlett Learning publications are available to corporations, professional associations, and other qualified organizations. For details and specific discount information, contact the special sales department at Jones & Bartlett Learning via the above contact information or send an email to specialsales@jblearning.com.

26872-0

Production Credits
VP, Product Development: Christine Emerton
Director, Product Management: William Larkin
Director, Content Management: Donna Gridley
Sales Manager: Brian Hendrickson
Project Manager: Kristen Rogers
Project Specialist: John Fuller
Senior Digital Project Specialist: Angela Dooley
Director of Marketing Operations: Brian Rooney
Production Services Manager: Colleen Lamy
VP, Manufacturing and Inventory Control: Therese Connell
Project Management: S4Carlisle Publishing Services
Composition: S4Carlisle Publishing Services
Cover Design: Kristin E. Parker
Text Design: Scott Moden
Rights Specialist: Liz Kincaid
Media Development Editor: Faith Brosnan
Cover Image (Title Page, Part Opener, Chapter Opener): © Jones & Bartlett Learning. Photographed by Glen E. Ellman.
Printing and Binding: LSC Communications
Cover Printing: LSC Communications

Library of Congress Cataloging-in-Publication Data
Library of Congress Cataloging-in-Publication Data
Names: Gunderson, Michael R., author. | Lindsey, Jeffrey, author.
Title: Applications of fire research and improvement / Michael Gunderson, Jeffrey Lindsey.
Description: Second edition. | Burlington, Massachusetts : Jones & Bartlett Learning, [2021] | Includes bibliographical references and index.
Identifiers: LCCN 2020023433 | ISBN 9781284206456 (paperback)
Subjects: LCSH: Fire protection engineering--Research. | Fire prevention--Research.
Classification: LCC TH9130 .G86 2021 | DDC 628.9/2072--dc23
LC record available at https://lccn.loc.gov/2020023433

6048

Printed in the United States of America

24 23 22 21 20   10 9 8 7 6 5 4 3 2 1

Brief Contents

Contents

About the Authors

Mic Gunderson, EMT-P (Ret.), FAEMS

Mic Gunderson has been in emergency services for over 40 years with third-service, fire rescue, and private and military emergency medical services (EMS) organizations. He is currently the president of the Center for Systems Improvement, a quality-improvement consulting and education firm that specializes in services to EMS, fire rescue, and healthcare organizations. He is an adjunct faculty member for the Master's in Emergency Health Services Degree Program at the Department of Emergency Health Services, University of Maryland–Baltimore County, where he has taught EMS system design, ambulance service contracting, and quality management. He was also the course developer for the EMS Quality Management Course at the National Fire Academy.

Mic has served on the editorial boards for several peer-reviewed EMS journals, including the *Journal of the American College of Emergency Physicians Open (JACEP Open), Prehospital and Disaster Medicine, Prehospital Emergency Care,* and the *EMS Management Journal.* Early in his career, he developed an interest in research working in a shock and resuscitation research laboratory at a medical school, which led him to start the first peer-reviewed EMS journal, which later evolved into *Prehospital and Disaster Medicine.* He served for 15 years on the faculty of the National Association of EMS Physicians' National EMS Medical Director's Course.

Mic is a frequent lecturer at EMS and other medical conferences across the United States and internationally and has had his work published extensively in both peer-reviewed and trade journals.

Jeffrey T. Lindsey, PhD, EMT-P, CHS IV, EFO, CFO

Dr. Lindsey is the chief learning officer for the Health Safety Institute. He is the program director of Fire and Emergency Services at the University of Florida. He retired from the fire service as the fire chief for Estero Fire Rescue in Estero, Florida. He has authored a number of textbooks for Brady Publishing.

He is an experienced leader, educator, lecturer, author, and consultant in emergency services. Dr. Lindsey earned his doctorate and master's degree in curriculum and instruction from the University of South Florida (USF). He holds a bachelor's degree in Fire and Safety Engineering from the University of Cincinnati and an associate degree in paramedic from Harrisburg Area Community College. He also has earned his Chief Fire Officer designation and completed the Executive Fire Officer Program.

Dr. Lindsey has been involved in the emergency services industry since 1980. He has a diverse background, with experience as a paramedic, fire fighter, 911 dispatcher, and educator. He served in various leadership roles, including fire chief.

He was a past member of Pre-hospital Research Forum, representative to the Fire and Emergency Services Higher Education EMS Degree Committee, and liaison for the International Association of Fire Chiefs (IAFC) EMS section to the American College of Emergency Physicians (ACEP) and has held a number of other appointments throughout his career. He served on the inaugural National EMS Advisory Council and was the safety chairperson of the council. He is a past member of the State of Florida EMS Advisory Council, vice chair of the Florida Fire Chief's EMS section, National Association of Emergency Medical Technicians (NAEMT) Governor representing Florida, board member of the National Association of EMS Educators (NAEMSE), and chairperson of the Strategic Planning Committee of the Lee County Fire Chiefs' Association.

CHAPTER

Introduction to Research

LEARNING OBJECTIVES

Upon completion of this chapter, you should be able to:

- Consider what research is and why we study it.
- Describe fire-related research objectives.
- Describe research and its foundations.
- Demonstrate an understanding of the scientific method.
- Identify the key elements of the agenda in the United States for fire-related research.

Case Study

The Somewhereville Fire Department was considering switching to a new type of hose as part of ongoing efforts to improve fire-ground operations. A hose manufacturer claims it has a new type of hose that is lighter and stronger—and therefore faster to deploy.

Before committing to a purchase that would replace much of the existing hose supply, which was still in good condition, the department got the manufacturer to agree to let the department test the new type of hose on its training grounds. The department's quality chief collaborated with the operations chief to design a study to evaluate the new hose in a scientifically sound manner.

The hypothesis of the study was that the new type of hose would allow crews to deploy faster compared with the existing type of hose. The department would determine this by studying the time from arrival on the scene to the first water on the fire. At the training grounds, the department had several sets of crews perform the same evolution with the same truck, the same structure, the same fire location, and so forth. One set of evolutions was performed with the existing type of hose, and the other set of evolutions used the new type of hose.

Seven different crews each ran three evolutions with the existing hoses and three with the new hoses, in random order. The on-scene times and the times for the first water on the fire were recorded. The data were then analyzed to compare the intervals between on-scene times and the times for first water on the fire for the evolutions with the existing hose to the evolutions with the new hose. No significant differences were found in the time intervals between the two types of hoses.

Despite the manufacturer's claims that the hose could be deployed faster, this was not what the department's testing showed. As a result, the purchase was not made. This saved the department thousands of dollars by preventing the purchase of equipment that was presumed to make a significant improvement in their fire-ground operations.

1. How does your department currently decide on new equipment purchasing decisions?
2. Do members of your department currently have skills in the design of experiments and data analysis to conduct a study similar to this case study?

Introduction

Much of the energy that propels the wheels of progress comes from research. In simple terms, basic **research** seeks ways to better understand what there is and how and why it does what it does. Applied research tries to apply that knowledge to solve everyday problems.

Basic fire research explores the chemistry, properties, and behavior of fire. Closely related are *applied* research efforts, which seek ways to prevent ignition, retard its spread, detect its presence, and enhance its extinguishment. Most of this type of research is done on laboratory benches or in the carefully controlled and monitored conditions of fire research laboratories at universities or manufacturing firms. We do not anticipate that many of you who are reading this text will be engaged directly in this type of fire research. However, we do hope that you will be educated users of such research. We hope that after reading this text, alone or in conjunction with a fire research course, you will have the ability to understand and critically evaluate peer-reviewed fire science research (**FIGURE 1-1**). As a fire service professional, you may choose to participate in efforts to influence legislation,

FIGURE 1-1 Fire departments can continue operating as they have always done, or the fire service can build a body of research to improve the effectiveness, efficiency, and safety of the service.

regulations, and policymaking at a local, state, or national level. This may require fire service professionals to use the results of fire laboratory research to guide such initiatives.

However, we *do* anticipate that most of you are, or will become, fire officers in a position to write and implement policies and procedures and make equipment and technology purchases. We hope to give you the knowledge and insights you will need to look at such changes as hypotheses that may improve the performance of specific fire department processes. From that perspective, we hope to help you learn how to apply the principles of research and performance improvement to design process-improvement experiments, test hypotheses, evaluate results, and make adjustments to optimize how well and how efficiently those processes meet the needs of your department and its stakeholders.

The Scientific Method

The scientific method is the basis for research in fire operations. It involves developing and testing a hypothesis. A *hypothesis* is simply an idea for a solution to a problem or a potential answer to a question. The potential solution or answer is tested in an experiment. Consider the following scenario.

You are attending a fire trade show and see a new type of hose in the exhibition hall. Or perhaps you just read an advertisement for the new type of hose in a fire magazine. The sales pitch sounds compelling. The new type of hose is said to be lighter, stronger, more flexible—and comes in a very cool new color. It suggests that this hose will make your department the envy of all the other departments in your area. You want your department to have all the latest technology because, after all, you want your department to be the best. You have equipment upgrade funds available to make the purchase.

Commonly, this type of purchase is made, the hose is put into service, and you get feedback from the crews. Some may be favorable, particularly from firefighters who like anything new. They rally around the notion that the latest technology is always the best technology. You may also get negative feedback, particularly from firefighters who embrace tradition and want things to stay the same. They rally around the notion that if it isn't broken, why fix it? Which group is right?

In approaching this issue from a scientific perspective, you should be thinking in terms of the dependent and independent variables affecting the type of hose you use. A *dependent variable* is a measure of how well the process under consideration is working. In this case, the process is fire suppression, and the dependent variable might be how long it takes to get water on a fire after a crew arrives on the scene. An *independent variable* is something you think might influence the dependent variable. In this case, the independent variable is the type of hose you use on your primary preconnect line. The sales pitch implies that by using this new type of hose, you can get water on the fire faster. Your hypothesis is "Will using this new type of hose (the independent variable) change the time interval between arriving at the scene and first water on the fire (the dependent variable)?"

How might you test whether the new hose gets water on the fire faster than the hose you are using now? You might leave the old hose on some trucks and put the new hose on other trucks. You could then compare the times between arrival on the scene and first water on the fire. But what if the crews on trucks with the old hose use more "hustle" in getting the job done compared to the other crews? What if the trucks using the new hose happen to run into a couple of hydrants with valve-opening problems that the other trucks do not? What if the fire is farther from the truck on the calls to which the crews using the new hose respond? These are examples of extraneous variables. An *extraneous variable* is something that might change the dependent variable (time interval from arrival on scene to first water on fire) in a way that has nothing to do with the independent variable (the type of hose).

To minimize the number or impact of any extraneous variables, we can try to control some of the circumstances. Instead of an actual fire, we might compare the interval from arrival on scene to first water on fire using the two different types of hoses in evolutions on a fire training ground. We could have the same distance from the hydrant to the fire, the same truck, the same crew, the same location of the fire in the burning building. We can control these factors in the design of our study to minimize their impact on the dependent variable. These factors that we can control are called, logically enough, *controlled variables*. The only thing that we want to change between fire evolutions is the independent variable—which is what type of hose is loaded on the truck.

There are also other factors to consider:

- Does it matter which type of hose is tested first?
- Should you run the evolution just once or several times for each type of hose?

- If you run the evolution more than once, how many times should you run it?
- If you run the evolution more than once, should you run all the evolutions on the same day?

We will explore these and other issues in more detail later on. At this point, we hope you are asking yourself whether your department makes scientific evaluations regarding changes to your equipment and standard operating procedures, similar to the department described in the Case Study. What about other policies and procedures? Think about it. More often than not, a change in policy, procedure, protocol, equipment, medications, techniques, or technology is a hypothesis. The change you make is usually being made because you believe that the change will make a particular process work better. Far too often, however, such changes are made with inadequate assessment—if any at all. Assessments should be made using the sound application of research and process-improvement techniques, as this book is designed show you.

The National Fire Service Research Agenda

What types of things should the fire industry as a whole and individual fire departments be looking at through research? In 2015 a group of fire service professionals asked themselves that very question. They held a symposium to update the national fire service research agenda. The purpose of the symposium was to produce an updated document that would identify and prioritize the areas in which research efforts should be carried out to support improvements in fire-fighter life safety. The resulting document, the National Fire Service Research Agenda (NFSRA), illustrates several types of research that you may find yourself getting involved with or reading about. Some of this research could be conducted at the level of the individual fire department. You may find the **16 Firefighter Life Safety Initiatives** (Appendix A) and the listing of recommendations (Appendix B) from the NFSRA to be useful starting points for coming up with your own research project ideas.

Wrap-Up

CHAPTER SUMMARY

The emphasis on research in the fire service is escalating to a higher degree than ever before. The U.S. Fire Administration has established research as a major priority as part of its mission. This text will provide the foundation for individuals in the fire service to gain an understanding of research, both how to conduct research and how to apply research that has been conducted.

KEY TERMS

16 Firefighter Life Safety Initiatives A set of key strategies that must be implemented to meet the U.S. Fire Administration's goal of providing a safer work environment for fire service personnel.

Research Scholarly or scientific investigation or inquiry.

REVIEW QUESTIONS

1. Select one of the 16 Firefighter Life Safety Initiatives and describe how research could affect the ability to implement or achieve this initiative.
2. Describe the process of the scientific method.
3. Using the scientific method, create an example to illustrate how the method works.
4. Explain why research is an important part of the fire service.
5. What constitutes research?
6. Describe fire-related research objectives.
7. Describe research and its foundations as it pertains to the fire service.
8. Identify the key elements of the agenda in the United States for fire-related research.
9. Explain how research benefits the fire service.
10. How would you overcome negative reactions in your department toward research?

ADDITIONAL RESOURCES

Leedy, Paul D., and Jeanne Ellis Ormond. 2019. *Practical Research: Planning and Design.* 12th ed. New York: Pearson Education, Inc.

Mahoney, E. 2009. *Mathematics and Problem Solving for Fire Service Personnel.* Upper Saddle River, NJ: Pearson.

National Fire Protection Association: Research Foundation, https://www.nfpa.org/News-and-Research/Resources/Fire-Protection-Research-Foundation.

National Institute of Standards and Technology: Fire Research Division, https://www.nist.gov/el/fire-research-division-73300.

National Institute of Standards and Technology: Online Database for the Building and Fire Research Laboratory Library (FIREDOC EXPRESS), https://www.nist.gov/el/firedoc.

Underwriters Laboratory: Fire Equipment Services, https://www.ul.com/offerings/fire-equipment-testing-service.

University of Maryland Department of Fire Protection Engineering: Fire Protection Research Page, https://fpe.umd.edu/research.

U.S. Fire Administration, http://www.usfa.dhs.gov.

CHAPTER 

Reviewing the Literature

LEARNING OBJECTIVES

Upon completion of this chapter, you should be able to:

- Discuss the reasons for conducting a literature search.
- Conduct a literature search.
- Identify the proper method for conducting a literature search using keywords.
- Explain how to evaluate and critique published literature.

Case Study

The U.S. Department of Homeland Security's Science and Technology Directorate and the National Institute of Standards and Technology (NIST) sponsored a series of experiments in a masonry educational building to examine the ability of fire service positive-pressure ventilation (PPV) fans to limit smoke spread or remove smoke in areas where potential occupants might be located. PPV fans do this by creating pressure that is higher than that of the fire to control where smoke and hot gases flow in the building. The study was conducted and published by NIST as *Evaluating Positive Pressure Ventilation in Large Structures: School Pressure and Fire Experiments* (Kerber 2007). Preliminary experiments examined the pressure increase created by portable fans and mounted fans in different configurations and locations. The two main fire scenarios included a long hallway with classrooms and a gymnasium. Both scenarios included fires that produced a large amount of smoke and hot gases. Instrumentation was placed to assess tenability criteria and how PPV tactics can either increase or decrease tenability. Measurements included temperature, pressure, thermal imaging, and video views. In the limited series of experiments in the long hallways of this masonry educational building, the use of PPV to increase pressure to reduce temperatures, limit smoke spread, and increase visibility was effective. This series of experiments demonstrated that fire service PPV fans can be used successfully in large structures to increase the tenability of potential victims and improve conditions for firefighting crews. The full report can be downloaded from the NIST website at https://www.nist.gov/publications/evaluating-positive-pressure-ventilation-large-structures-school-pressure-and-fire.

1. Does your fire department access reports of this nature when considering ways to improve its strategies and tactics?
2. Does your department have expertise in conducting literature searches to find these types of reports?

Introduction

A **literature review** can be extremely helpful to a fire fighter or a fire officer trying to answer a question or solve a problem. It should be one of the first steps in almost every research or performance-improvement project. Why? Someone may have already done the work of answering the question or solving the problem you are considering, or something very similar to it. For this reason, a literature review is often required by organizations as part of their internal proposals for process-improvement projects or by funding agencies for research-grant applications. Even if the project you are considering is not the same as other previous studies, there may be an extensive body of literature that can help refine your questions or give you ideas about how to conduct your project. There are a variety of reasons for spending the time and effort to conduct a review of the literature before embarking on a project. A careful literature review can allow you to do the following (Bourner 1996):

- Find gaps in the literature.
- Avoid "reinventing the wheel" (at the very least, a literature review can save you time, and it may help you avoid making the same mistakes as others have before you).
- Build on the platform of existing knowledge and ideas.
- Learn about other people working in the same field (a professional network is a very valuable resource to cultivate).
- Identify important projects relevant to your topic.
- Provide the intellectual context for your own work, enabling you to position your project relative to other work—how it is similar, how it adds, how it differs.
- Learn about opposing views.
- Discover information and ideas that may be relevant to your project.
- Identify research methods that may be helpful to your project.

Literature Sources

There are many places to look for literature that may be pertinent to your question or project, including books,

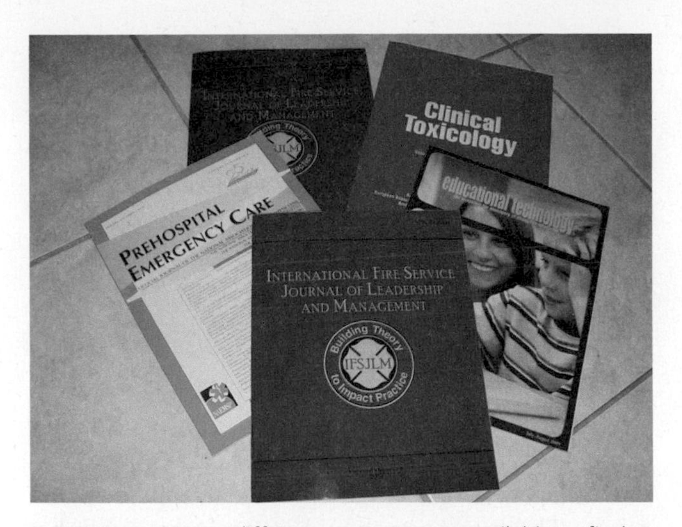

FIGURE 2-1 Many different resources are available to find literature on a subject in which you are interested.

journals, newspapers, government publications, conference proceedings, and websites (**FIGURE 2-1**).

A good way to start a literature search is to identify one or more **keywords** or short phrases relating to the topic of your question or project. These can be entered into a **search engine** to find relevant information on the World Wide Web. You probably have already used this sort of strategy when using a search engine such as Google. To choose keywords and phrases, look at your research problem. It should help you identify keywords and phrases to begin your search. For example, if you want to study the time savings and risks associated with using lights and sirens versus responding in a nonemergency mode to incidents, you might identify keywords such as *emergency response, sirens, emergency lights, emergency vehicles,* and *emergency driving.* The number of responses that you find during your search of the various types of literature may unveil thousands of results or "hits" related to these key terms. Some of the hits may not be relevant to your study. You will want to refine your search to reduce the number of nonrelevant hits. Once you have identified your keywords, you will want to look at various resources to begin to identify the literature that relates to your study. There are three major starting points for resources: online databases, paper catalogs, and indexes in libraries.

Online Databases

Computer database searches are the most efficient way to conduct a literature search today. The reach of the popular generic web search engines such as Google is vast, but they do not index everything. There are many specialized literature databases that may be helpful as well. Some of these can be accessed without charge; others charge a fee per search or require a subscription. Searching these latter databases without charge is often possible by going through a college or public library. If you are a registered student, you may be able to get an account that allows you to access these databases from home or work by logging into your school's computer system and then conducting your searches through that portal to avoid fees. If you are not a student, many college and university libraries allow the public to use their library computer workstations on-site without a fee. If you are doing medically related searches, local hospitals often have libraries and computer workstations that you may be able to ask for permission to use. Your emergency medical services (EMS) medical director may be able to help you make these arrangements.

Search-String Construction

The keywords and phrases you enter into the search box of a generic or specialized search engine is called a **search string**. Your search string should be broad enough to produce as much relevant information as possible but specific enough to exclude as of much the nonrelevant information as possible. Most search engines are able to understand more than just the words in a search string. They can often understand special characters and commands to make the search more specific. The following information, adapted from Google's "How to Search on Google" (Google 2020a) and "Refine Web Searches" (Google 2020b) pages, is generally true for other search engines as well, but consult the help file or operating instructions for each search engine or database for the exact commands and syntax requirements for search strings.

Note: Words that are actually entered into the search engine for the examples shown below are in *italics.*

Every Word Matters

Generally, all the words you put in the query will be used. There are some exceptions. Searching is always case-insensitive. Searching for *new york times* is the same as searching for *New York Times.* With some exceptions, punctuation is ignored (that is, you can't search for @, #, $, %, ^, &, *, (), =, +, [], \, and other special characters).

Keep It Simple

If you're looking for a particular company, just enter its name or as much of its name as you can recall. If you're looking for a particular concept, place, or product, start with its name. If you're looking for a pizza restaurant, just enter *pizza* and the name of your town or your Zip Code. Most queries do not require advanced operators or unusual syntax. Simple is good.

Think how the page you are looking for will be written. A search engine is not a human; it is a program that matches the words you give to pages on the web. Use the words that are most likely to appear on the page. For example, instead of saying *my head hurts,* say *headache* because that's the term a medical page will use. The query *in what country are bats considered an omen of good luck?* is very clear to a person, but the document that gives the answer may not have those words. Instead, use the query *bats are considered good luck in,* or even just *bats good luck*, because that is probably what the right page will say.

Describe what you need using as few terms as possible. The goal of each word in a query is to focus it further. Because all words are used in the search, each additional word limits the results. If you limit the search too much, you will miss a lot of useful information. The main advantage to starting with fewer keywords is that if you don't get what you need, the results will likely give you a good indication of what additional words you need to include to refine your results on the next search. For example, *weather cancun* is a simple way to find the weather, and it is likely to give better results than the longer *weather report for cancun mexico.*

Choose descriptive words. The more descriptive the word is, the more likely you are to get relevant results. Words that are not very descriptive, such as *document, website, company,* or *info,* are usually not needed or very useful. Keep in mind, however, that even if the word has the correct meaning, but it is not the one that most people use, it may not appear on the pages you need. For example, *celebrity ringtones* is more descriptive and specific than *celebrity sounds.*

Phrase Search (" ")

If you put double quotation marks around a set of words, you are telling Google to consider the exact words in that exact order. Google already uses the word order and the fact that the words are together as a very strong signal and will search another way only for a good reason, so quotes are usually unnecessary. By insisting on phrase search, you might be missing good results accidentally. For example, a search for *"Alexander Bell"* (with quotes) will miss the pages that refer to Alexander G. Bell.

Search within a Specific Website (site:)

Google allows you to specify that your search results must come from a specific website. For example, the query *iraq site:nytimes.com* will return pages about Iraq, but it will only include those from nytimes.com. A simpler query, *iraq nytimes.com* or *iraq New York Times,* will usually be just as good, although it might return results from other sites that mention the *New York Times.* You can also specify a whole class of sites; for example, *iraq site:.gov* will return results only from a .gov domain, and *iraq site:.iq* will return results only from Iraqi sites.

Terms You Want to Exclude (-)

Attaching a hyphen (-), which functions as a minus sign, immediately before a word indicates that you do *not* want pages that contain this word to appear in your results. The minus sign should appear immediately before the word and should be preceded by a space. For example, in the query *anti-virus software,* the minus sign is used as a hyphen and will not be interpreted as an exclusion symbol, whereas the query *anti-virus -software* will search for the word "anti-virus" but exclude references to software. You can exclude as many words as you want by using the minus sign in front of all of them, for example, *jaguar -cars -football -os.* The minus sign can be used to exclude more than just words. For example, place a hyphen before the "site:" operator (without a space) to exclude a specific site from your search results.

Fill in the Blanks (*)

The asterisk symbol (*), or wildcard symbol, is a little-known feature that can be very powerful. If you include * within a query, it tells Google to treat the star as a placeholder for any unknown term(s) and then find the best matches. For example, the search *Google ** will give you results about many of Google's products. The query *Obama voted * on the * bill* will give you stories about different votes on different bills. Note that the * operator works only on whole words, not parts of words.

Search Exactly as Is (+)

Google employs synonyms automatically, so it finds pages that mention, for example, childcare for the query *child care* (with a space) or California history for the query *ca history*. But sometimes Google helps out a little too much and gives you a synonym when you don't really want it. By attaching a plus sign (+) immediately before a word (remember, don't add a space after the +), you are telling Google to match that word precisely as you typed it. Putting double quotation marks around the word will do the same thing.

The OR Operator

Google's default behavior is to consider all the words in a search. If you want to specifically allow either one of several words, you can use the OR operator (note that you have to type *OR* in ALL CAPS). For example, *San Francisco Giants 2004 OR 2005* will give you results about either one of these years, whereas *San Francisco Giants 2004 2005* (without the *OR*) will show pages that include both years on the same page. The symbol | can be substituted for OR. (The AND operator, by the way, is the default, so it is not needed.)

Exceptions

Searching is rarely absolute. Search engines use a variety of techniques to imitate how people think and to approximate their behavior. As a result, most rules have exceptions. For example, the query *for better or for worse* will not be interpreted by Google as an OR query but as a phrase that matches a (very popular) comic strip. Google will show calculator results for the query *34 * 87* rather than use the "fill in the blanks" operator. Both cases follow the obvious intent of the query. The following sections note some exceptions to some of the rules and guidelines that were mentioned in this discussion and the Basic Search Help article.

Exceptions to "Every Word Matters." Words that are very commonly used, such as *the*, *a*, and *for*, are usually ignored (these are called *stop words*). However, there are even exceptions to this exception. A search for *the who* likely refers to the band; the query *who* probably refers to the World Health Organization—Google will not ignore the word *the* in the first query.

Synonyms might replace some words in your original query. (Adding + before a word disables synonyms.)

A particular word might not appear on a page in your results if there is sufficient other evidence that the page is relevant. The evidence might come from language analysis that Google has done or many other sources. For example, the query *overhead view of the bellagio pool* will give you nice overhead pictures from pages that do not include the word *overhead*.

Punctuation That Is Not Ignored. Punctuation in popular terms that have particular meanings, like *C++* or *C#* (both are names of programming languages), are not ignored.

The dollar sign ($) is used to indicate prices. *nikon 400* and *nikon $400* will give different results.

The hyphen (-) is sometimes used as a signal that the two words on either side of it are very strongly connected (unless there is a space before the hyphen but no space after it, in which case it is interpreted as a minus sign).

The underscore symbol (_) is not ignored when it connects two words, for example, *quick_sort*.

Google and other search engines often have "advanced search" windows that take care of some of the details of search-string construction by letting you fill out a form with appropriate information.

FIGURE 2-2 shows the results of a Google search for information on the effect of using lights and sirens on response time. Google looked for information using the following three phrases: *emergency lights;*

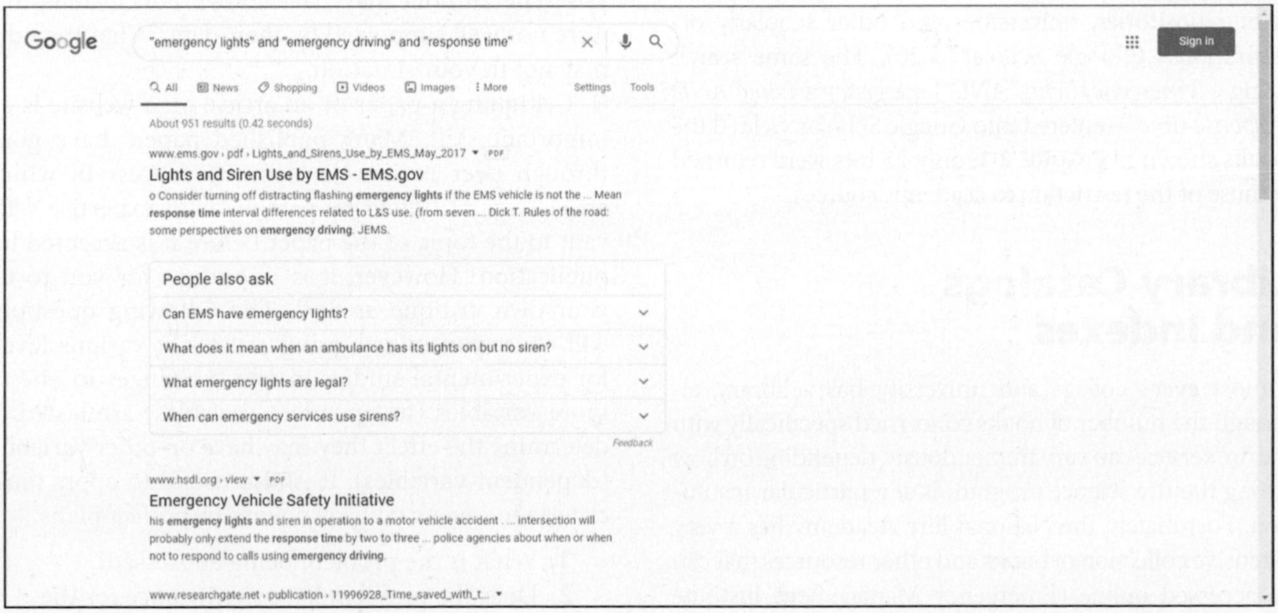

FIGURE 2-2 Google search results for information about the effect on response times when driving with emergency lights on.

Courtesy of Google, Inc.

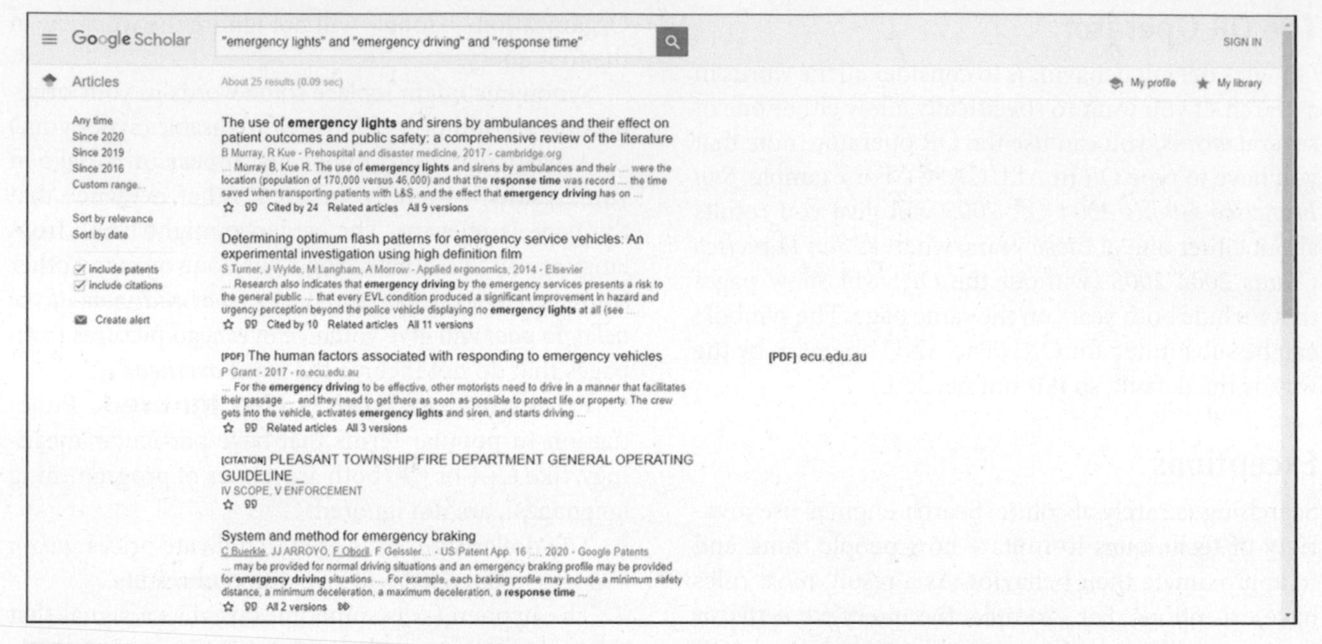

FIGURE 2-3 Google Scholar results on the effect on response time when driving with emergency lights on.

emergency driving; response time. Each phrase was set inside quotation marks and was separated from other search terms by the word AND. Thus, the search string entered into the search engine was *"emergency lights" AND "emergency driving" AND "response time"*. This search returned 951 different sources or "hits."

Google also offers a specialized search engine called Google Scholar. It "provides a simple way to broadly search for scholarly literature. From one place, you can search across many disciplines and sources: peer-reviewed papers, theses, books, abstracts and articles, from academic publishers, professional societies, preprint repositories, universities and other scholarly organizations" (Google Scholar 2020). The same search string—*"emergency lights" AND "emergency driving" AND "response time"*—entered into Google Scholar yielded the results shown in **FIGURE 2-3**; only 25 hits were returned because of the restriction to academic sources.

Library Catalogs and Indexes

Almost every college and university has a library, although the number of books concerned specifically with the fire service can vary tremendously, depending on how strong the fire science program is at a particular institution. Fortunately, the National Fire Academy has a very extensive collection of books and other resources that can be accessed online (Emergency Management Institute n.d.). This collection includes research papers by participants in the Executive Fire Officer Program—perhaps the largest collection of community department-level fire science research literature. The National Fire Academy library participates in a national library program that allows people to check out books from the National Fire Academy through their local library.

Critical Review

You should apply a critical eye when reviewing books, articles, and other literature. Some of what you find may have used a poor study design, methods, or analysis. The authors may have drawn conclusions that were not well supported by their data. What they did may not fit your situation.

Critiquing a paper or an article on a website is an important skill. Many published papers have gone through peer review—a rigorous process in which articles are critiqued by persons with expertise relevant to the topic of the paper before it is accepted for publication. However, it is important for you to do your own critique as well. The following questions will assist you in critiquing articles at various levels for experimental studies in which changes to one or more variables (independent variables) are tested to determine the effect they may have on other variables (dependent variables). It is also useful to apply these criteria to review your own work or project plans.

1. What is the problem being addressed?
2. Does the problem have significance for the fire service in general or to your specific department?

3. Is the problem expressed as a purpose, research question, or hypothesis to be tested?
4. Are the purpose statements or research questions worded so that key concepts or variables are clearly identified? Is the study group or population studied clearly specified?
5. Is the study design appropriate for answering the research question or achieving the purpose of the study?
6. What procedures, if any, did the researcher use to control external (situational) factors and intrinsic (subject characteristic) factors? Were these procedures appropriate and adequate?
7. How were data actually recorded (e.g., field notes, checklists)? Did the recording procedure appear appropriate?
8. What was the plan by which events or behaviors were sampled? Did this plan appear appropriate?
9. Who collected the research data? Were the data collectors qualified for their role, or is there something about them (e.g., their professional role, their relationship with study participants) that could undermine the collection of unbiased, high-quality data?
10. How were data collectors trained? Does the training appear to have been adequate?
11. Where and under what circumstances were the data gathered? Were other people present during the data collection? Could the presence of others have created any distortions?
12. Did the collection of data place any undue burdens (in terms of time or stress) on participants? How might this have affected data quality?
13. Does the report include any descriptive statistics? Do these statistics sufficiently describe the major characteristics of the researcher's data set?
14. Were appropriate descriptive statistics used (e.g., were percentages reported when a mean would have been more informative)?
15. Were groups compared to determine the statistical significance of any differences found between groups?
16. Was a statistical test performed for each of the hypotheses or research questions?
17. Were the statistical tests appropriate (e.g., were the tests appropriate for the level of measurement of key variables)?
18. Were the results of any statistical tests significant? Nonsignificant? What do the tests tell you about the plausibility of the research hypotheses?
19. Was an appropriate amount of statistical information reported? Were important analyses omitted, or were unimportant analyses included?
20. Is the researcher objective in reporting the results?
21. Does the research have the potential to help solve a problem faced by the fire service?

Wrap-Up

CHAPTER SUMMARY

Conducting a literature search can be a challenge. It is important to be able to define your search using keywords so as to return the most pertinent documents. In addition, using a proper search string may make the difference in locating the resources you are trying to find. Once you conduct your search and have your documents, it is then important to be able to critique the documents to sort out those that are relevant and put aside those that are not. Conducting a literature search is a critical step in the research process.

KEY TERMS

Keywords Words used to search for literature pertaining to a certain topic or area of knowledge.

Literature review Used to find previous published research on a topic of interest. It is a means to read about others' findings on a particular subject.

Search engine A website that allows the user to enter keywords to locate documents or other forms of literature on the Internet or in a database or system.

Search string Keywords and phrases that are entered into the search box of a generic or specialized search engine.

REVIEW QUESTIONS

1. You are writing an article on smoke detectors. You want to find additional information on smoke detectors and what research on smoke detectors has been published. How would you go about performing an electronic literature search to locate valid research on this topic?
2. What are the reasons for conducting a literature search?
3. Select a research article from a peer-reviewed source. Perform a critical review of the article using the criteria in this chapter.
4. List the keywords you would use to conduct a literature search on the prevention of fire-related deaths of children.
5. Explain how you can evaluate published literature.
6. Describe how you would critique and select the appropriate published research.
7. Why is it important that the literature be credible?
8. What criteria would you use to perform your evaluation of literature?
9. Describe creating a search string.

ADDITIONAL RESOURCES

Google Scholar—for academically oriented searches, http://scholar.google.com.

Google Search—general-purpose searches, http://www.google.com.

U.S. Fire Administration (USFA) Research Projects—USFA initiatives currently under way in the areas of fire detection, suppression, and notification systems; first responder health and safety; home electrical wiring; and others, http://https://www.usfa.fema.gov/data/library/research/topics/.

REFERENCES

Bourner, T. 1996. "The Research Process: Four Steps to Success." In *Research Methods: Guidance for Postgraduates*, edited by T. Greenfield. London: Arnold.

Emergency Management Institute. n.d. Accessed June 9, 2020, https://training.fema.gov/emi.aspx.

Google. 2020a. "How to Search on Google." Accessed June 11, 2020, http://www.google.com/support/websearch/bin/answer.py?answer=134479.

Google. 2020b. "Refine Web Searches." Accessed June 11, 2020, http://www.google.com/support/websearch/bin/answer.py?hl=en&answer=136861.

Google Scholar. 2020. "About." Accessed June 11, 2020, http://scholar.google.com/intl/en/scholar/about.html.

Kerber, Steven, Daniel Madrzykowski, and David Stroup. 2007. *Evaluating Positive Pressure Ventilation in Large Structures: School Pressure and Fire Experiments* (NIST Interagency/Internal Report [NISTIR] 7412 (March). Accessed June 11, 2020, https://nvlpubs.nist.gov/nistpubs/Legacy/IR/nistir7412.pdf.

CHAPTER

Performance Management

LEARNING OBJECTIVES

Upon completion of this chapter, you should be able to:

- Explain the three broad phases of quality improvement for an organization.
- Identify the various accrediting agencies in the emergency services industry.
- Discuss process improvement as it relates to the performance management of a fire department.

Case Study

The Somewhereville Fire Department, as part of its long-term quality-management strategy, decided to begin its quality journey with a review of compliance with regulatory requirements and standards. The reviewers found that they were at an excellent level of compliance, but they still made a few adjustments to make their compliance rock solid and reliable on an ongoing basis. That process was completed in just a few months, so they quickly moved to the next phase, deciding to move in parallel on two fronts. They decided to begin using the Criteria for Performance Excellence from the Malcolm Baldrige National Quality Program to guide their long-term performance-improvement efforts. In addition, they would begin the process of self-assessment using the fire department accreditation criteria from the Center for Public Safety Excellence® (CPSE®).

For the Baldrige program, their first-year activities consisted of completing their organizational profile. They sent two of their officers to a Baldrige training sponsored by their state Baldrige association. The chief of the department attended an executive-level Baldrige program orientation course, along with other executives from across the state, that was also sponsored by the state Baldrige association.

For the fire service accreditation track, the department established a project team consisting of several officers and a few frontline fire fighters to begin the assessment process and make recommendations to the command staff.

By the end of that first year, the chief, along with the two officers who attended the Baldrige training, were able to lay out a general plan for the next 3 years on how they would be working their way up through the various levels of their state Baldrige association's program. Also at the end of the first year, the accreditation program provided specific recommendations on what to address in preparation for submitting the department's accreditation application.

1. Has your department ever conducted a compliance audit?
2. Does your state have a Baldrige program?
3. Has your department ever applied for accreditation from the CPSE®? If yes, what issues were found that needed correction? If no, what were the reasons for not applying for accreditation?
4. If your department provides ambulance service, has it ever applied for accreditation from the Commission on Accreditation of Ambulance Services? If yes, what issues were found that needed correction? If no, what were the reasons for not applying for accreditation?

Introduction

Organizations are complex. They can be compared to a living being, with an anatomy and a physiology. The anatomy includes intricate operational, financial, and social systems and physical structures. These are often documented by organizational charts, reporting hierarchies, and articles of incorporation. The organizational physiology consists of the many levels of processes, subprocesses, and sub-subprocesses that continue to lower and lower levels until you reach the smallest possible process details. These are typically documented by policies, procedures, and protocols.

When someone is sick, treatment may be directed toward improving the state of the entire body (e.g., better nutrition, exercise, or removal from a toxic environment) or may focus on a particular part or function (e.g., surgical repair of damaged parts or medication to stabilize or correct a particular physiological process). Similarly, improving the performance of an organization can be approached by improving the overall organization or by focusing on a specific process.

Organizational Improvement

It is helpful to think of improvement at an organizational or process level as a journey rather than a destination (City of Coral Springs, Florida n.d.). The organizational-improvement journey can be approached in three broad phases that are generally addressed in the following order: a regulatory phase, an accreditation phase, and a benchmarking phase.

The Regulatory Phase

The **regulatory phase** of organizational improvement seeks to ensure that all applicable regulations, laws, contractual requirements, and so forth are being complied with. Many organizations have a **compliance officer** whose primary responsibility is to ensure compliance with applicable local, state, and federal laws and regulations and third-party guidelines; manage audits and investigations into regulatory and compliance issues; and respond to requests for information from regulatory bodies (Wikipedia n.d.).

Your fire department's organizational-improvement journey should begin with compiling a comprehensive inventory of the requirements that apply to your department. These may include regulations of local, state, and federal government agencies, such as the U.S. Department of Transportation. These agencies work together to create standards of performance on multiple levels to help build public confidence. The next step is to audit your department to see if it is in compliance, and then take steps to get into compliance if any areas are found to be lacking. The final step is to develop and carry out an ongoing process to monitor the requirements for any changes and to take steps to ensure continued compliance.

The Accreditation Phase

Once your department has completed the regulatory phase and is in a monitoring mode with periodic reaudits, it is ready to move into the **accreditation phase**. Of the several definitions of *accreditation* offered by the Merriam-Webster Online Dictionary (n.d.), the one most applicable to the improvement journey is "to recognize or vouch for as conforming to a standard."

Professional associations within an industry often establish **minimum standards** that all professionals or organizations in the industry should be in compliance with to be considered competent. Following a prescribed process to document and verify compliance with those standards is an accreditation process. The accreditation standards are generally not intended to identify the most exemplary individuals or organizations in the industry; they are intended to provide a standard that all organizations should be able to achieve with reasonable effort. By establishing such standards, the level of performance for the entire industry can be elevated as all participants achieve compliance with the accreditation standards. This is commonly seen in other business sectors as well, such as in universities and hospitals.

The accreditation phase begins with compiling a comprehensive inventory of all accreditation standards that apply to the department overall or to specific processes or functions. This is followed by efforts to work toward compliance with those standards.

FIGURE 3-1 Fire departments that meet the performance indicators set forth by the CFAI® will be recognized as being nationally accredited.

Courtesy of the Wilson Fire Rescue Services.

At the level of the overall fire department, the accredidation model developed by the **Commission on Fire Accreditation International® (CFAI®)** is the most applicable (**FIGURE 3-1**). CFAI® accreditation is "an all-hazard, quality improvement model based on risk analysis and self-assessment that promotes the establishment of community-adopted performance targets for fire and emergency service agencies" (CPSE n.d.-b). CFAI® is part of the Center for Public Safety Excellence® (CPSE®) "an international technical organization that works with the most progressive fire and emergency service agencies and most active fire professionals" and whose mission it is to "lead the fire and emergency service to excellence through the continuous quality improvement process of accreditation, credentialing, and education" (CPSE n.d.-a).

Many local government executives are pressured to justify any increase in expenditures unless they are attributed directly to improved or expanded service delivery in the community. More than ever before, these local leaders are faced with the constant pressure of doing more work with less funding. The CFAI® model is developed by city/county management, fire chiefs, labor, and other subject matter experts and guides fire and EMS personnel to assess professional performance and efficiency of an organization's achievement. The CFAI® model guides organizations in developing benchmarks and provides tangible data and information for elected officials.

In addition to the CFAI® standards, fire departments should identify other applicable accreditation standards. The **National Fire Protection Association (NFPA)** offers voluntary standards. Although these are not accreditation standards per se, they do represent consensus standards that departments can work toward.

FIGURE 3-2 EMS agencies that meet the CAAS standards display this seal on their units.

Courtesy of City of Henderson Fire Department.

If the department is involved in emergency medical services (EMS), the **Commission on Accreditation of Ambulance Services (CAAS)** (**FIGURE 3-2**) offers applicable standards. Some of those standards may not be applicable for departments that are not involved in transport, but other parts of the CAAS standards may still apply to fire first-response programs. A 911 communications center that handles emergency calls and dispatches fire apparatus may look toward standards compliance or accreditation from the **Commission on Accreditation for Law Enforcement Agencies (CALEA)**, the **Association of Public-Safety Communications Professionals (APCO)** (APCO n.d.), and the **National Academies of Emergency Dispatch (NAED)** (NAED n.d.).

Fire departments are often part of larger city or county government entities. Accreditation standards for these cities or counties may include standards for fire departments or emergency services departments. For example, the American Public Works Association (APWA) has accreditation standards for emergency management (APWA n.d.).

An issue to consider when looking at how to approach the accreditation phase is whether to use a self-assessment or external verification with formal accreditation. For a self-assessment, your department would obtain copies of the accreditation standards and manuals. Your staff would use them to evaluate whether your department meets each of the accreditation standards. Any shortcoming would be identified and corrected until your department is in full compliance. Self-assessment has the advantage of avoiding any submission fees and external examiner travel expenses that may be associated with formal external accreditation reviews.

However, there are significant disadvantages to a purely internal accreditation standards assessment process:

- Internal staff may not be entirely objective, particularly in addressing deficiencies.
- Internal staff usually have less training and practice in interpreting and applying standards—in contrast to external reviewers, one of whom will be a "lead" reviewer who has experience with several prior organizational reviews.
- There will be no external verification that your department actually meets the standards.
- There will be no opportunity to celebrate the achievement of formal accreditation.

Once formal accreditation is obtained or internally determined compliance is achieved, the performance-improvement journey is just getting started. Unfortunately, many organizations fail to recognize the limitations of accreditation in their journey. They behave as if achieving accreditation means they have "arrived" at organizational performance excellence. Accreditation and other industry standards allow a department to identify gaps in performance associated with being *competent*. They do not identify the superlative characteristics of excellence as found in the best-performing organizations. Meeting accreditation standards is like clearing a hurdle at a track meet: It is a pass–fail proposition. Clearing the hurdle means only that you jumped high enough to pass the test. No extra points are awarded for jumping higher than required. Indeed, hurdles are *not* designed to measure how high you actually jumped—or how much higher you jumped over the hurdle this time compared with previous times. The benchmarking phase thus takes up where the accreditation phase leaves off.

The Benchmarking Phase

In the **benchmarking phase**, your department can begin to transcend the standards it worked so hard to meet during the regulatory and accreditation phases. Those standards were the minimums; now it is time for the organization to work toward achieving even higher levels of performance. However, without further regulatory and accreditation standards to seek compliance with, what targets does your department shoot for? That is where benchmarking comes in.

With internal benchmarking, your organization compares its current performance to its own past performance. The general idea is to find ways to do things better today than you did yesterday. Over time, these small incremental changes can build into substantial positive differences. External benchmarking, on the other hand, offers the opportunity for your department to compare its level of performance to other organizations. The purpose of external benchmarking is to learn from what others have achieved and then incorporate the best parts of their methods into your organization. Getting that external perspective can bring ideas to light that your department may never have considered with internal benchmarking.

At the organizational level, the challenge is to find a tool that can be used to measure performance and facilitate external benchmarking comparisons. Accreditation measures are not designed for that. As mentioned, those standards are pass–fail. They tell you whether the organization exceeded or failed the standard, but nothing more. An entirely different type of tool is needed to measure overall organizational performance on a scale. The tool should allow repeat measurements that can assess progress over time. The tool should not be prescriptive, specifying how performance is to be improved, as standards commonly are. Organizations should have creative latitude in finding ways to improve their performance.

There is one category of assessment tools that is uniquely positioned to meet this need. These are the criteria used in various states and countries for their performance-recognition programs. The concept began in Japan with the criteria for the Deming Prize (W. Edwards Deming Institute n.d.) and was adopted by the United States to create the **Baldrige National Quality Program** and its Criteria for Performance Excellence (National Institute of Standards and Technology, Baldrige Performance Excellence Program n.d.-a). Similar programs have been established in other countries (e.g., Canada Awards for Excellence, National Quality Institute, and European Foundation for Quality Management Award).

The Baldrige Criteria for Performance Excellence are a useful tool for fire departments to utilize in measuring their overall level of performance:

> Whether your organization is small or large; is involved in service, manufacturing, government, or nonprofit work; and has one office or multiple sites across the globe, the Criteria provide a valuable framework that can help you measure performance and plan in an uncertain environment. The Criteria can help you align resources with approaches such as ISO 9000, Lean, a Balanced Scorecard, and Six Sigma; improve communication, productivity, and effectiveness; and achieve strategic goals. (National Institute of Standards and Technology, Baldrige National Quality Program, Criteria for Performance Excellence, www.quality.nist.gov/Business_Criteria.htm, accessed April 7, 2009.

As early as practical in your department's performance-improvement journey, consider getting involved in the Baldrige evaluation process. The earlier your department can complete a Baldrige evaluation, the sooner it will have a baseline from which progress can be objectively measured. Most states have Baldrige programs that include training and support resources (The Baldrige Alliance n.d.).

The national-level Baldrige program provides a way for fire departments to get started doing self-assessments using a simple online questionnaire. It provides a snapshot of your organization, the key influences on how you operate, and the key challenges you face. The first section, "Organizational Description," addresses your organization's operating environment and your key relationships with customers, suppliers, partners, and stakeholders. The second section, "Organizational Challenges," calls for a description of your organization's competitive environment, your key strategic challenges and advantages, and your system for performance improvement. If you identify topics for which conflicting, little, or no information is available, it is possible that the "Organizational Profile" can serve as your complete assessment, and you can use these topics for action planning (National Institute of Standards and Technology, Baldridge Performance Excellence Program n.d.-b).

The national Baldrige program also offers a publication called *Getting Started* (National Institute of Standards and Technology, Baldridge Performance Excellence Program n.d.-c), which describes the benefits of self-assessment and the many tools, approaches, and resources available to help you get started using the Baldrige Criteria. You can use these tools and approaches to convince others in your organization of the value of conducting a Baldrige self-assessment.

Another early-stage Baldrige self-assessment toolset is called "Are We Making Progress?" (National Institute of Standards and Technology, Baldridge Performance Excellence Program 2015). Deploying your organization's strategy can be much more difficult than developing it. This easy-to-use questionnaire can help you assess how your organization is performing and learn what can be improved. Based on the Baldrige Criteria for Performance Excellence, the questionnaire will help you focus your improvement and communication efforts on areas needing the most attention from the perspective of your employees. There is a national database of results from this survey that can reveal how your employees' perceptions compare with the perceptions of employees of other organizations in all sectors—business, education, health care, and nonprofit—and can help your organization set priorities for improvement. "Are We Making Progress as Leaders?" is a companion piece to the "Are We Making Progress?" tool that is specifically for senior managers. It asks the same questions as the "Are We Making Progress?" questionnaire, but from the perspective of leadership. These questionnaires are provided for use by any organization, at no cost, to help in organizational-improvement efforts.

After your department has taken and utilized the feedback from these entry-level self-assessment questionnaires, it will be ready to start using the full Baldrige Criteria assessment process. There are three versions—for business/not-for-profit, education, and healthcare organizations. The government/not-for-profit version (National Institute of Standards and Technology, Baldridge Performance Excellence Program n.d.-e) may be the most applicable to fire departments with regard to their mission of fire prevention and suppression. The EMS division of a fire department can use the healthcare criteria (National Institute of Standards and Technology, Baldridge Performance Excellence Program n.d.-d) separately, if desired.

All three versions have the same general structure, as shown in **FIGURE 3-3**.

The Organizational Profile (at the top of the figure) sets the context for the way your organization operates. Your environment, key working relationships, and strategic challenges and advantages serve as an overarching guide for your organizational performance management system.

The System Operations are composed of the six Baldrige categories in the center of the figure that define your operations and the results you achieve. Leadership (Category 1), Strategic Planning (Category 2), and Customer Focus (Category 3) represent the leadership triad. These categories are placed together to emphasize the importance of a leadership focus on strategy and customers. Senior leaders set your organizational direction and seek future opportunities for your organization.

Workforce Focus (Category 5), Process Management (Category 6), and Results (Category 7) represent the results triad. Your organization's workforce and key processes accomplish the work of the organization that yields your overall performance results. All actions point toward the Results category—a composite of product, customer, market, financial, and internal operational performance results, including workforce, leadership, governance, and societal responsibility results.

The horizontal arrow in the center of the framework links the leadership triad to the results triad—a linkage that is critical to organizational success. Furthermore, the arrow indicates the central relationship between Leadership (Category 1) and Results (Category 7). The two-headed arrows indicate the importance of feedback in an effective performance management system.

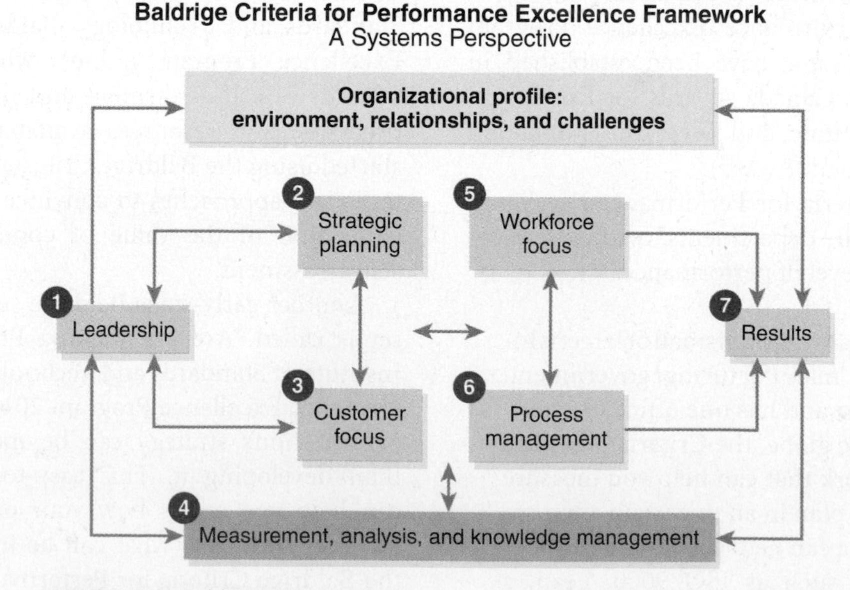

FIGURE 3-3 Baldrige Criteria structure.

National Institute of Standards and Technology, Baldrige National Quality Program, Criteria for Performance Excellence, www.quality.nist .gov/Business_Criteria.htm, accessed April 7, 2009.

The System Foundation area is represented by the box at the bottom of the figure. Measurement, Analysis, and Knowledge Management (Category 4) are critical to the effective management of your organization and to a fact-based, knowledge-driven system for improving performance and competitiveness. Measurement, analysis, and knowledge management serve as a foundation for the performance management system.

More information and resources pertaining to the use of the Baldrige process for assessing overall organizational performance are provided in the Additional Resources section at the end of this chapter.

Process Improvement

Process improvement, in the context of fire service, is a scientific/data-driven approach to monitoring and improving performance in a particular process. It can be used to solve problems, reduce errors, improve efficiency, improve quality, reduce delays, and better satisfy the needs of those the process serves, either directly or indirectly.

There are many ways to improve a process. Unfortunately, many fire service professionals do not use a process-improvement strategy per se; management intuition seems to be more common. Veteran officers draw on their years of experience and training to come up with what they feel is the best way to address a problem or meet a need. Sometimes, several officers and line firefighters may be given an opportunity to participate in the discussions and debates that lead to a new policy, procedure, protocol, equipment choice, or implementation of new technology. Hopefully, a consensus will emerge that considers all of the pros and cons and leads to a decision that everyone supports. In that regard, the consensus building may have been quite successful, but what about the decision itself?

The problem is that this intuitive/expert approach to process improvement usually does not include an explicit definition of the problem or issue to be addressed or an objective assessment of how the expert- or consensus-based solution actually changed the performance of the process under consideration. Opinions that express how well the changes are working may be solicited from or offered spontaneously by members of the department. However, that is a far cry from solid data that can be used in an objective analysis of how well the process was working before the change in contrast with how well it did afterward.

There is a lesson for local fire departments to learn from the researchers who conduct studies on the science of fire behavior, building materials and construction engineering to prevent or inhibit fires, and chemical sciences on retarding and extinguishing fires. These professional disciplines apply the principles of the scientific method to evaluate their hypotheses of what does or does not make an improvement. Those results drive their conclusions and catalyze improvements in legislation, building codes, and chemical agents for firefighting.

The mainstream manufacturing and service industries, as well as many sectors of government, have been incorporating the principles of the scientific method into their approach to process improvement for decades. Many EMS operations within fire departments have also been using contemporary process-improvement methods for quite some time. Unfortunately, most EMS process-improvement programs have had very limited success because of their use of conceptually flawed models imported from earlier models used in health care that were based on misguided attempts to reduce the risk of financial losses in settlements and litigation from errors and bad outcomes rather than actual process improvement (Berwick, Godfrey, and Roessner 1990). Process-improvement "best practices" are just as applicable to the EMS aspects of fire department activities as they are to operations involving the fire ground and support.

The collective experiences in the manufacturing, service, and government sectors of mainstream industry offer fire service professionals a rich body of literature on process improvement to learn from. Quality planning, quality assurance, quality improvement, Kaizen (a Japanese strategy for continuous improvement), Total Quality Management, Six Sigma, lean, and error proofing are some of the topic names you can find on this subject in public and university libraries, in local bookstores, and on the web. The body of literature on performance improvement specifically for fire operations is small (see, e.g., Bruegman 2002; Coleman 1999; Evans and Dyar 2009; Janing and Sachs 2003; Wallace 2001), but hopefully it will expand over time as these principles and techniques are incorporated into the ways that community fire departments approach their problems, challenges, and opportunities for improvement. Process improvement in local fire departments can be accomplished using the **DMAIC** (Define—Measure—Analyze—Improve—Control) framework of the Six Sigma methodology (ISixSigma n.d.), as detailed in the later chapters of this book.

THE RESEARCH EXPERIENCE

The United States alone spends about $700 billion per year on new and renovated construction. About 20 percent of this is to ensure safety from unwanted fires, which includes the cost of insurance, to make families whole after fires and to recover from business loss. This is an enormous cost to endure every year. Combined with a growing construction market in other countries, this presents a major opportunity for the introduction of new fire-safe products to the building and transportation industries and new products, such as advanced detectors and suppression systems, and firefighting equipment for the fire protection industry. The intent of performance-based standards is to provide flexibility in maintaining accepted fire safety from unwanted fires with new competitive products while providing an opportunity for saving lives and reducing property loss, at the same time gaining a reduction in the cost of design, construction, maintenance, and liability coverage.

In order to derive this benefit, it is necessary to have tools to evaluate building system performance, which then provide a metric for the effectiveness of design and material use. The methodology embodied in HAZARD I is intended as a tool to aid in understanding the consequences of unwanted fires. The intention of the HAZARD I methodology is to make available the research that is done in pursuit of this goal. Improvements will include increased applicability of the current procedures, improved usability, the ability to address additional building features, and more accurate treatment of the fire itself and the effects of the fire on people and their actions. Many improvements have been made in the documentation that accompanies the software as it has evolved. These improvements are a result of the experience fire protection engineers and others have had in using the methodology. The experience gained by having HAZARD I widely used, concomitant with the improvements that are now being incorporated, constitutes the first step in the overall goal of creating a complete fire hazard assessment methodology.

Jones, W. 1997. "The Evolution of HAZARD, the Fire Hazard Assessment Methodology." *Fire Technology* 33: 167–182.

Wrap-Up

CHAPTER SUMMARY

Improving the performance of an organization can be approached by improving the overall organization or by focusing on a specific process within it. The organizational-improvement journey can be approached in three broad phases that are generally addressed in the following order: a regulatory phase, an accreditation phase, and a benchmarking phase. Process improvement can be used to solve problems, reduce errors, improve efficiency, improve quality, reduce delays, and better satisfy the needs of those the process serves, either directly or indirectly. These processes can help improve an organization and should be considered in every organization.

KEY TERMS

Accreditation phase The phase of organizational improvement in which an independent agency recognizes or affirms that an organization conforms to an established standard.

Association of Public-Safety Communications Professionals (APCO) An organization that sets standards for 911 emergency agencies.

Baldrige National Quality Program A set of criteria used for performance recognition.

Benchmarking phase The phase of organizational improvement in which an organization begins to transcend the standards it worked so hard to comply with during the regulatory and accreditation phases.

Commission on Accreditation for Law Enforcement Agencies (CALEA) An organization that establishes standards for law enforcement agencies.

Commission on Accreditation of Ambulance Services (CAAS) An organization that sets standards for ambulance agencies.

Commission on Fire Accreditation International® (CFAI®) An organization that establishes standards for fire departments.

Compliance officer Someone whose primary responsibility is to ensure compliance with applicable local, state, and federal laws, regulations, and third-party guidelines and who manages audits and investigations into regulatory and compliance issues and responds to requests for information from regulatory bodies.

DMAIC Define—Measure—Analyze—Improve—Control: a framework used in the Six Sigma methodology.

Minimum standards Standards with which all professionals or organizations in the industry should be in compliance to be considered competent.

National Academies of Emergency Dispatch (NAED) An organization that sets standards for 911 emergency agencies.

National Fire Protection Association (NFPA) An organization that sets standards on fire-related topics.

Process improvement A scientific/data-driven approach to monitoring and improving performance in a particular process.

Regulatory phase The phase of organizational improvement that seeks to ensure that all applicable regulations, laws, contractual requirements, and so forth are being complied with.

REVIEW QUESTIONS

1. Describe the three phases of organizational improvement.
2. Select an accreditation agency. Research it and describe the requirements for obtaining accreditation from that agency.
3. Why should a fire department consider becoming accredited?
4. Explain the benefits of process improvement.
5. Explain process improvement as it relates to performance management in a fire department.
6. Explain the benchmarking phase and its importance for a fire department.
7. Describe the Baldrige Award and the benefits it has for a fire department.
8. Explain the role of a compliance officer.
9. What does DMAIC stand for?
10. What are the significant disadvantages of a purely internal process for the assessment of accreditation standards?

ADDITIONAL RESOURCES

American Society for Quality, Quality in Government Section, http://www.asq.org/government/why-quality/overview.html.

Association of Public-Safety Communications Officials. n.d. Accessed June 8, 2020, https://www.apcointl.org/standards/standards-to-download/.

Brown, M. G., *Baldrige Award Winning Quality*, 17th ed. New York: Taylor & Francis, 2008.

Canadian National Quality Institute, Quality Criteria for the Public Sector, http://www.asq.org/government/why-quality/overview.html.

Center for Public Safety Excellence, https://cpse.org.

Clark, K., *Baldrige Self-Assessment: Practical Tips for Increasing the Value Add from the Process*. Milwaukee, WI: American Society for Quality, 2006.

Leonard, D., and M. McGuire, *The Executive Guide to Understanding and Implementing the Baldrige Criteria: Improve Revenue and Create Organizational Excellence*. Milwaukee, WI: American Society for Quality, 2007.

REFERENCES

American Public Works Association. n.d. Accessed June 8, 2020, https://www.apwa.net/MYAPWA/Education_Credentialing/Accreditation/MyApwa/Apwa_Public/Education_and_Events/Agency_Accreditation.aspx.

Berwick, D. B., A. B. Godfrey, and J. Roessner. 1990. *Curing Healthcare*. San Francisco: Jossey-Bass.

Bruegamn, R. 2002. *Exceeding Customer Expectations: Quality Concepts for the Fire Service*. Upper Saddle River, NJ: Prentice Hall.

Center for Public Safety and Excellence®. n.d.-a "CPSE Overview." Accessed August 3, 2020. https://cpse.org/cpse-overview/.

Center for Public Safety and Excellence®. n.d.-b Home page. Accessed August 3, 2020. https://cpse.org/.

City of Coral Springs, Florida. 2008. "Coral Springs Is First City to Win the Baldridge Award." *Coral Springs* (Winter). Accessed June 8, 2020, https://www.coralsprings.org/home/showdocument?id=1524.

Coleman, R. 1999. "Benchmarks: Your Tool for Winning Wars, Not Battles." *Fire Chief* 43, no. 9: 32–33.

Evans, B., and J. Dyar. 2009. *Management of EMS*. Upper Saddle River, NJ: Brady.

International Academies for Emergency Dispatch. n.d. "Accreditation." Accessed June 8, 2020, https://www.emergencydispatch.org/Accreditation.

ISixSigma. n.d. "What Is Six Sigma?" Accessed June 11, 2020, https://www.isixsigma.com/new-to-six-sigma/getting-started/what-six-sigma.

Janing, J., and G. Sachs. 2003. *Achieving Excellence in the Fire Service*. Upper Saddle River, NJ: Prentice Hall.

Jones, W. 1997. "The Evolution of HAZARD, the Fire Hazard Assessment Methodology." *Fire Technology* 33: 167–182.

Merriam-Webster Online Dictionary. n.d. "Accredit." Accessed June 11, 2020, http://www.merriam-webster.com/dictionary/accredit.

National Institute of Standards and Technology, Baldridge Performance Excellence Program. 2015. "Are We Making Progress?" Accessed June 8, 2020, https://www.nist.gov/baldrige/self-assessing/improvement-tools/are-we-making-progress.

National Institute of Standards and Technology, Baldrige Performance Excellence Program. n.d.-a. Accessed June 8, 2020, https://www.nist.gov/baldrige.

National Institute of Standards and Technology, Baldridge Performance Excellence Program. n.d.-b. "Baldridge Organizational Profile." Accessed June 8, 2020, https://www.nist.gov/baldrige/baldrige-organizational-profile.

National Institute of Standards and Technology, Baldridge Performance Excellence Program. n.d.-c. "Getting Started with Baldridge." Accessed June 8, 2020, https://www.nist.gov/baldrige/self-assessing/getting-started.

National Institute of Standards and Technology, Baldridge Performance Excellence Program. n.d.-d. "Baldrige by Sector: Health Care." Updated November 15, 2019. Accessed June 8, 2020, https://www.nist.gov/baldrige/self-assessing/baldrige-sector/health-care.

National Institute of Standards and Technology, Baldridge Performance Excellence Program. n.d.-e. "Baldrige by Sector: Nonprofit." Updated November 15, 2019. Accessed June 8, 2020, https://www.nist.gov/baldrige/self-assessing/baldrige-sector/nonprofit/government.

The Baldridge Alliance. n.d. "Why the Baldridge Alliance?" Accessed June 8, 2020, https://www.baldrigealliance.org/.

W. Edwards Deming Institute. Accessed June 11, 2020. http://www.deming.org.

Wallace, Michael J. 2001. *Benchmarking Fireground Performance: Executive Development* (November 1). (National Fire Academy Executive Fire Officer Program Applied Research Project Paper). Accessed June 8, 2020, https://www.hsdl.org/?view&did=4867.

Wikipedia. n.d. "Chief Compliance Officer." Last updated August 2019. Accessed June 8, 2020, http://en.wikipedia.org/wiki/Chief_Compliance_Officer.

CHAPTER

Defining and Measuring the Problem

LEARNING OBJECTIVES

Upon completion of this chapter, you should be able to:

- Describe how to define the problem.
- Discuss the various ways of measuring a problem.
- Identify various models used in measuring.

Case Study

The Somewhereville Fire Department noticed that it was having problems with more frequent residential structure fires. The fire marshal reported that the number of fires had increased every year for the past 3 years. As a result, the department chief established an ad hoc improvement project team to address the problem. A project charter was written for the team. This defined the problem that the team would be focused on and set expectations for the deliverables and timelines for the team.

1. When your department conducts a major project, is a project team formally designated to carry out the project? Does it have a written document that clearly identifies the project team's purpose, expected deliverables, and timelines? If not, how are such projects managed? Has that approach been successful?
2. Consider the process of pump operations on the fire ground. Create a diagram that shows the suppliers, inputs, process, outputs, and customers (known as a SIPOC—pronounced "sigh-pock"—diagram) for the pump-operations process.

JONES & BARTLETT LEARNING NAVIGATE 2 *Access Navigate for more resources.*

Introduction

To begin your quest for answers to your research question, you need to be able to define the problem. Once you have defined the problem, you need to be able to measure the effect the problem has on your organization. This chapter looks at the process used in defining the problem and then discusses methods and models for measuring the problem.

Defining the Problem

The **define phase** of a project is about clarifying exactly what problem or issue you are trying to address. The best place to start is by looking at the problem or issue from the perspective of a customer.

Good research and good improvement projects usually come from careful consideration of your fire department's internal and external customers and what they really need or want. *Customers* are those individuals, groups, or organizations that utilize the output of your work processes. Internal customers are within your fire department. External customers are outside of your fire department. There is also the ultimate customer: the person, group, or organization for whom the process exists. In the case of fire services, this is usually the property owner and occupants for fire responses and patients for medical responses.

A very common error is to make assumptions about what your internal, external, or ultimate customers need or desire. Whenever possible, ask them! You should try to determine not only what they may need or want but also how they would like to get it, how often they want it, in what form they want it, what factors are more or less important to them, and so on. Explorations of customer needs and wants are called *voice-of-the-customer* (VOC) studies (iSixSigma n.d.; Wikipedia 2020). Excellent opportunities for research and improvement are found in the gaps between the processes customers need and desire versus what they currently receive.

Return on Investment

You will often find yourself faced with several viable ideas for ways to make improvements. Your bias should be toward selecting projects that have the greatest impact for the customer or that have the greatest potential to provide a significant **return on investment** (ROI). The ROI may be operational, financial, political, or clinical. Here are some examples of ROI:

- Operational
  - Faster response times
  - Shorter fire task times (e.g., on-scene to fire extinguished)
  - Fewer fires
  - Improved equipment reliability (reduced downtime)
  - Reduced hiring-cycle time (position vacant to position filled)
  - Fewer fire-ground tactical errors and omissions
- Financial
  - Lower costs per fleet mile
  - Lower costs per the following:
    - Recruit application received
    - Recruit applicant interviewed

- Recruit applicant hired
- New employee orientation completed

- Political
  - Increased satisfaction rates for the following:
    - Property owners
    - Patients
    - Employees
    - External public safety and healthcare team members
  - Decreased rates of external customer complaints
  - Improved reputation ratings
- Clinical
  - Improved success rates for medical procedures (e.g., airway management success rate)
  - Improved intermediate and discharge outcomes
    - Cardiac arrest
      - Increased return of spontaneous circulation rate
      - Increased rate of survival to hospital admission
      - Increased rate of survival to hospital discharge
    - S-T elevation myocardial infarctions (STEMIs) and strokes
      - Reduced interval from symptom onset to 911 notification
      - Reduced interval from 911 notification to balloon (STEMI) or stroke intervention
      - Decreased length of hospital stay
  - Decreased rates of medical errors

Once you have a good sense of what the issue is that you are going to address with your project, you should write it out as a problem statement. A well-expressed problem is specific, measurable, and manageable. It should not infer blame, guess at the cause, or guess at the solution. With these criteria in mind, consider the following examples:

"We have missed our 8:59 emergency response time standard 23% of the time in the last 6 months."

- Is it specific? Yes: It names a particular process and what the problem is.
- Is it measurable? Yes: Response time can be measured.
- Is it manageable? Yes: The problem is limited to one type of response.
- Does it infer blame? No.
- Does it guess at the cause? No.
- Does it guess at the solution? No.

"Our company's procedure for responding to calls has many delays due to communication problems."

- Is it specific? Yes: It identifies the process being considered.
- Is it measurable? Yes: Delays can be measured.
- Is it manageable? Yes.
- Does it infer blame? Yes: It blames communications.
- Does it guess at the cause? Yes: It is up to the project team to find the cause of a problem. Preconceived ideas about cause may be inaccurate, incomplete, or mistaken and can mislead the team. In this example, after collecting and analyzing data, it might turn out that routing and/or improper fleet management is the cause of the problem.
- Does it guess at the solution? No.

Measuring the Baseline

Armed with a good problem statement, you are ready to enter the **measure phase** of your project. This is when you make a detailed assessment of the current status of the process (or processes) involved. You will get to know, in very specific ways, how the process operates now. This will provide a **baseline** so that when changes are made later, you will have something to compare the new process performance against. It will also help you identify areas for potential improvements.

Various tools are commonly used during the measure phase of an improvement project. They include the following:

- Performance indicators
- Statistical process control charts
- Capability indexes
- Suppliers, Inputs, Process, Outputs, and Customers (SIPOC—pronounced "sigh-pock") tables
- Flowcharts
- Value stream maps

During the measure phase of your project, you will want to measure the current level of performance in your target process using **performance indicator(s)**. A performance indicator shows how well (quality) or how efficiently (economically) processes are performing (Gunderson 2009). Think of performance indicators as gauges that tell you the level of performance for a particular process. In research terms, performance indicators are dependent variables (recall Chapter 1, *Introduction to Research*). The value of the indicator *depends* on the quality or efficiency level of the process in question.

You will continue using these performance indicators throughout the rest of the project to see if the changes you hypothesize for making improvements during the **improve phase** of your project were effective. You will also use them for the long term, after the project, to monitor performance over time so that you can intervene if necessary if signs of performance deterioration appear.

The National Fire Protection Association's Fire Service Performance Measures (Flynn 2009) offer suggested indicators that include the following:

- Percentage of total fires responded to that spread beyond the room of origin after fire department arrival
- Number of civilian deaths (injuries)/1000 fires
- Number of fire fighter fatalities (injuries)/1000 fires
- Average dollars saved per fire
- Percentage of preventable fires
- Percentage of (fire) inspections not completed in target cycle

When you are designing your process performance indicator, you will want to make sure it provides enough detail to make operational implementation clear and consistent. It is also helpful to use the same or similar formats as other organizations with which you may want to make comparisons or exchange data (Gunderson 2009). An example of fire service performance indicator format is the **Fire Service Performance Indicator Format (FSPIF) model** shown in **TABLE 4-1**.

The application of the FSPIF is best illustrated by an example. **TABLE 4-2** shows a performance indicator definition for the rate at which low-hazard commercial property inspections are completed.

TABLE 4-1 Fire Service Performance Indicator Format (FSPIF)

- Indicator/Attribute Name: Name or title of the performance indicator.
- Key Process Path: Starting with one of the predefined key process names, this item shows which key process and subprocess the indicator reflects on.
- Customer/Need: Indicators are designed to reflect on how well and/or how efficiently a given customer need is being met. This item shows what customer and need the indicator reflects on.
- Type of Measure: Structure, process, or outcome.
- Objective: Describes why an indicator is useful in specifying and assessing the process or outcome of services measured by the indicator.
- Indicator/Attribute Formula: The equation for calculation of the indicator. If applicable, separate sections will separately address the numerator and denominator of the indicator equation.
- Indicator/Attribute Formula Description: Explanation of the formula used for the indicator. Where applicable, separate descriptions detailing the numerator and denominator will be provided.
- Denominator Description: Description of the population (group) being studied or other denominator characteristics, including any equation or other key aspects that characterize the denominator.
- Denominator Inclusion Criteria: Additional information not included in the denominator statement that details the parameters of the denominator population.
- Denominator Exclusion Criteria: Information describing criteria for removing cases from the denominator.
- Denominator Data Sources: Sources for data used in generating the denominator.
- Numerator Description: Description of the subset of the population being studied or other numerator characteristics, including any equation or other key aspects that characterize the numerator.
- Numerator Inclusion Criteria: Additional information not included in the numerator statement that details the parameters of the numerator population.
- Numerator Exclusion Criteria: Information describing criteria for removing cases from the numerator.
- Numerator Data Sources: Sources for data used in generating the numerator.
- Sampling Allowed: Indicates if sampling the study population is or is not allowed in the calculation of this indicator.
- Sampling Description: If sampling is allowed, this will describe the sampling process to be used for this indicator.
- Minimum Number of Data Points: Tells how many data points are needed, at a minimum, for calculation of this indicator.

- Suggested Reporting Format—Numerical: The suggested way in which the numerical results should be expressed (e.g., decimal minutes, percentages, ratios).
- Suggested Reporting Format—Graphical: The suggested way in which reports should be presented in graphical format (e.g., pie charts, statistical process control charts).
- Suggested Reporting Frequency: Time frame, number of successive cases, or other grouping strategies by which cases should be aggregated for calculating and reporting results.
- Testing: Indicates if a formal structured evaluation has been performed on the various scientific properties of the indicator, such as its reliability, validity, and degree of difficulty of data collection.
- Stratification: Indicates if stratification has been applied to the indicator.
- Stratification Options: Suggested stratification criteria for use with this indicator.
- Current Development Status: Describes the amount of work completed to date relative to the final implementation of the indicator.
- Additional Information: Further information regarding an indicator not addressed in other sections.
- References: Citations of works used for the development of the indicator.
- Contributors: Listing of persons or organizations used in development and refinements to this indicator.

Based on National EMS Performance Indicator Project/Open Source EMS Initiative Performance Indicator format.

TABLE 4-2 Low-Hazard Commercial Property Fire Inspection Rate

Indicator Name	Low-hazard commercial property fire inspection rate
Key Process Path	Prevention > fire inspection process
Patient/Customer Need	Community need: Inspections at sufficient frequency to reduce fires from causes that could be prevented by low-hazard commercial inspections and correction actions
Type of Measure	Process (SPO[a]); quality (QCV[b])
Objective	Reduce the frequency of low-hazard commercial property fires
Indicator Formula	Number of inspections completed in specified time interval
Indicator Formula Description	n/a
Denominator Description	Time frame (e.g., monthly, weekly, daily)
Denominator Inclusion Criteria	n/a
Denominator Exclusion Criteria	n/a
Denominator Data Sources	n/a
Numerator/Percentile Data Point Description	Number of inspections completed in the specified time frame
Numerator/Percentile Data Point Inclusion Criteria	All inspections completed for commercial properties designated as "low hazard" by department guidelines
Numerator/Percentile Data Point Exclusion Criteria	Reinspections of properties that had issues significant enough to require a reinspection to ensure compliance per department guidelines

(*Continues*)

TABLE 4-2 Low-Hazard Commercial Property Fire Inspection Rate (*Continued*)

Numerator/Percentile Data Point Data Sources	Fire inspector activity log
Sampling Allowed	No
Sampling Description	n/a
Minimum Number of Data Points	5
Suggested Reporting Format—Numerical	Inspections completed per specified time frame (e.g., monthly, weekly, daily)
Suggested Reporting Format—Graphical	Run chart; statistical process control chart (X-bar and R chart pair)
Suggested Reporting Frequency	Daily (if at least 5 days), weekly (if at least 5 weeks), monthly
Indicator Testing	Not yet completed
Stratification	Yes
Stratification Options	By individual inspector; by department (when aggregating between several departments)
Current Development Status	General use of the definition widely utilized; operational details of definition need testing
Additional Information	None
References	None
Contributors	None

a. Donabedian, A., Evaluating the Quality of Medical Care, *Milbank Memorial Fund Quarterly, 44*(1):166–203, 1966.
b. Gunderson, M., The EMS Value Quotient: Looking at the Combined Effects of Costs and Quality, *Journal of Emergency Medical Services (JEMS)*, January 2009.

There are many details to work out in putting the performance indicator into operation, including how the data will be collected, how often the data will be analyzed, how the data will be reported, and how the results and raw data will be stored. For now, however, let's assume that those issues have been worked out and that you are getting daily results from your performance indicator.

Suppose that the fire marshal is facing budget cutbacks that will leave one vacancy in the civilian fire inspection staff unfilled. As a result of the unfilled vacancy, the fire marshal needs to find ways to improve the productivity of the remaining inspectors. The fire marshal wants to maintain the pace of inspections to meet the departmental goals of making at least one inspection of every low-hazard commercial property every 2 years. One of the key indicators of how well this process is working is the number of inspections that are completed during a given period of time. Suppose your community has 5000 businesses that fall into this category, and your department has a policy of conducting an inspection of these types of businesses at least once every 2 years. That means your department will need to have the low-hazard commercial property inspection process operating at a pace as described in **TABLE 4-3**.

So how often should you measure your inspection rate? Every second, every year? Or somewhere in between? You should seek a balance that gives you information about your process performance often enough to detect changes (good or bad) but not so often that there is too much information to utilize.

Considering the information in Table 4-3, measuring the inspection rate every second or every minute will clearly not be helpful because if the process is operating at anywhere near the target pace, you will only complete an inspection around once an hour. But trying to assess how well the process is working on an hourly basis, with only a single data point coming in per hour, doesn't make much sense either. Daily, you would expect to see around 10 inspections, so that

TABLE 4-3 Calculation of Desired Pace for Completing Low-Risk Commercial Property Inspections

Time Frame	Desired Pace of Inspections	Notes
Every 2 Years	5000	
Annually	2500	
Monthly	208.33	
Weekly	50	Excluding 2 weeks/yr for holidays
Daily	10	Weekdays only
Hourly	1.25	8-hour workday
Minute	0.0208	
Second	0.00035	

might be the most frequent interval that is practical. Coming at the question from the other direction, if you only measure the inspection rate every 2 years, you will not be able to detect any changes that occur in the process very quickly. The same goes for annually. If you measure monthly, it will take a few months to see whether any changes you make have an impact because you won't want to draw conclusions from just one data point. Thus, an assessment that occurs more often is probably needed. That gets you to weekly, which is probably the least-often interval you will want to consider. So, you can reasonably choose between weekly or daily measurement intervals.

Let's suppose you choose to calculate your performance indicator on a daily basis. Your challenge then becomes how to interpret the variation that you will see in the results from day to day. Every process has variation, and there are two types that you will need to understand and be able to differentiate: common-cause and special-cause variation.

Common-cause variation is what happens when the process is working as designed and without any special (unusual) circumstances acting on it in a way that significantly alters the performance-indicator results. Special-cause variation is what you see when special circumstances are affecting the results. This is a very important differentiation to make because improving a process with special-cause variation should be approached very differently from improving a process that shows only common-cause variation. In fact, applying improvement methods suited to removing undesirable special-cause variation can often damage the performance of a process that has only common-cause variation—and vice versa.

For example, if your fire department is like most, emergency response times are one of the performance indicators it monitors routinely. One part of that process that can be influenced is the turnout time (the time from crew notification to apparatus en route). Even if we look only at calls when the crew is in-station when the alarm is received, there will always be some variation in turnout time. If we take the average turnout time each day for an entire month and plot these averages on a chart, we might see results like those shown in the *run chart* in **FIGURE 4-1**. A run chart is a simple line graph that plots the value of a performance indicator or other measurement over time.

Notice that there are a few days when the turnout times are longer than others. Some managers might be inclined to bring this to the attention of the crews involved in those calls. The well-intentioned manager might tell the crews that their turnout time on those occasions was longer than on other days that month. The well-intentioned manager might take more aggressive actions if the same crews are involved in these slower responses again. But what if there isn't any significant difference between these slower responses and the rest of the responses? After all, in any group of responses, there will *always* be faster ones and slower ones. If there are no significant differences between the slower and faster responses, then there is no need to counsel crews about the slower responses. They are doing their job with consistency. And unwarranted counseling might cause the crews, knowing they are following policies and procedures with appropriate regard, to start doing things in a way that damages the integrity of the process. For example, they might start reporting that they are en route before they actually are, in an effort to avoid getting called out for it again and thereby damaging the validity of the turnout-time data. This is an example of treating a problem with common-cause variation as if it were a special-cause problem.

Now consider a similar scenario in which the well-intentioned manager sees that some turnout times are longer than others and counsels everybody to do better, when in fact, the longer turnout times are consistently coming from a single station. Again, dysfunctional behaviors may start to crop up among many of the crews. The specific problem at the specific station goes undetected and unresolved. This is an example of applying a common-cause strategy to a special-cause problem.

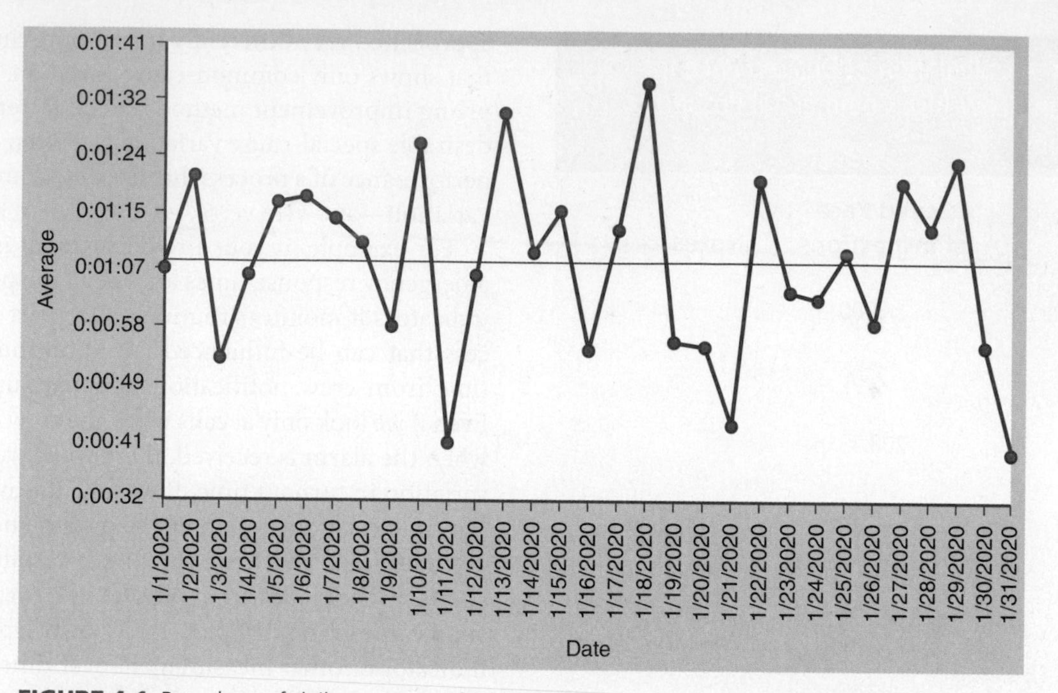

FIGURE 4-1 Run chart of daily averages of turnout times for a 1-month period.

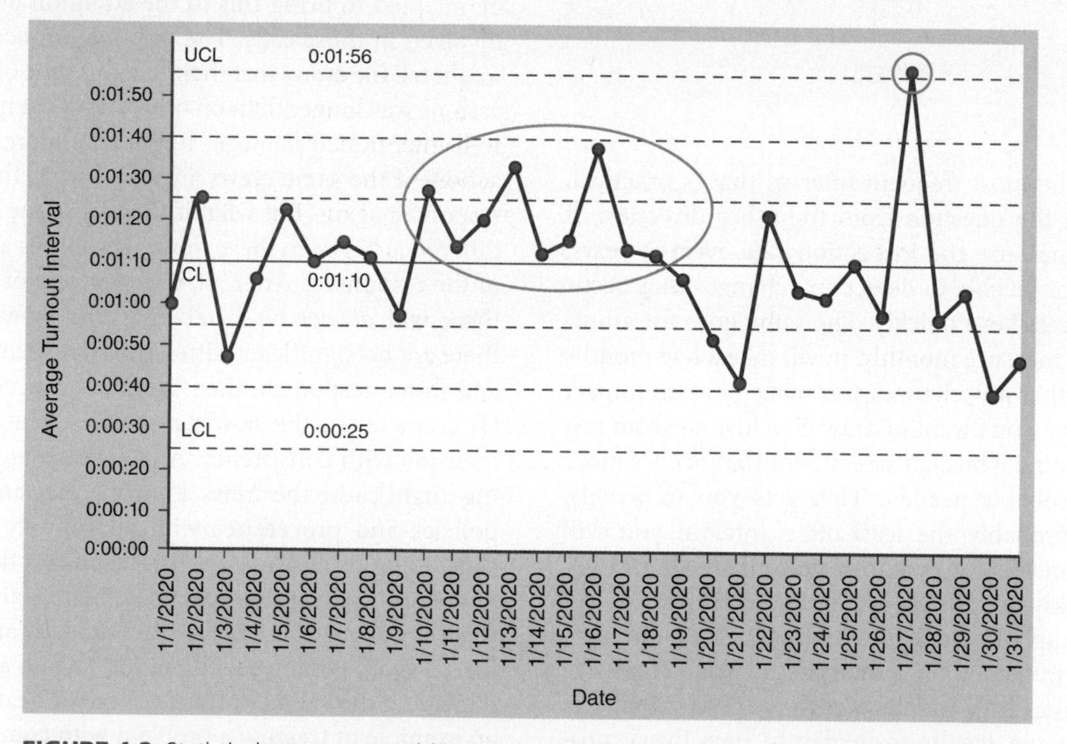

FIGURE 4-2 Statistical process control (SPC) chart of turnout-time data with a point above the upper control limit (UCL) and a set of eight points above the centerline.

Fortunately, there is a tool that is designed to differentiate between common-cause and special-cause variation: It is called a *statistical process control* (SPC) chart. Using our 1-month turnout-time data, software for generating SPC charts can identify the expected range into which values will fall when only common causes of variation are present. The upper and lower limits of that expected range of common-cause variation are called the *upper control limit* (UCL) and the *lower control limit* (LCL). There are also certain patterns of data that indicate special-cause variation is taking place, even if the data are falling within the control limits. However, when the data fall inside the control limits and there are no special-cause data patterns, the process is most likely to be operating with only common-cause variation. **FIGURE 4-2** shows an SPC

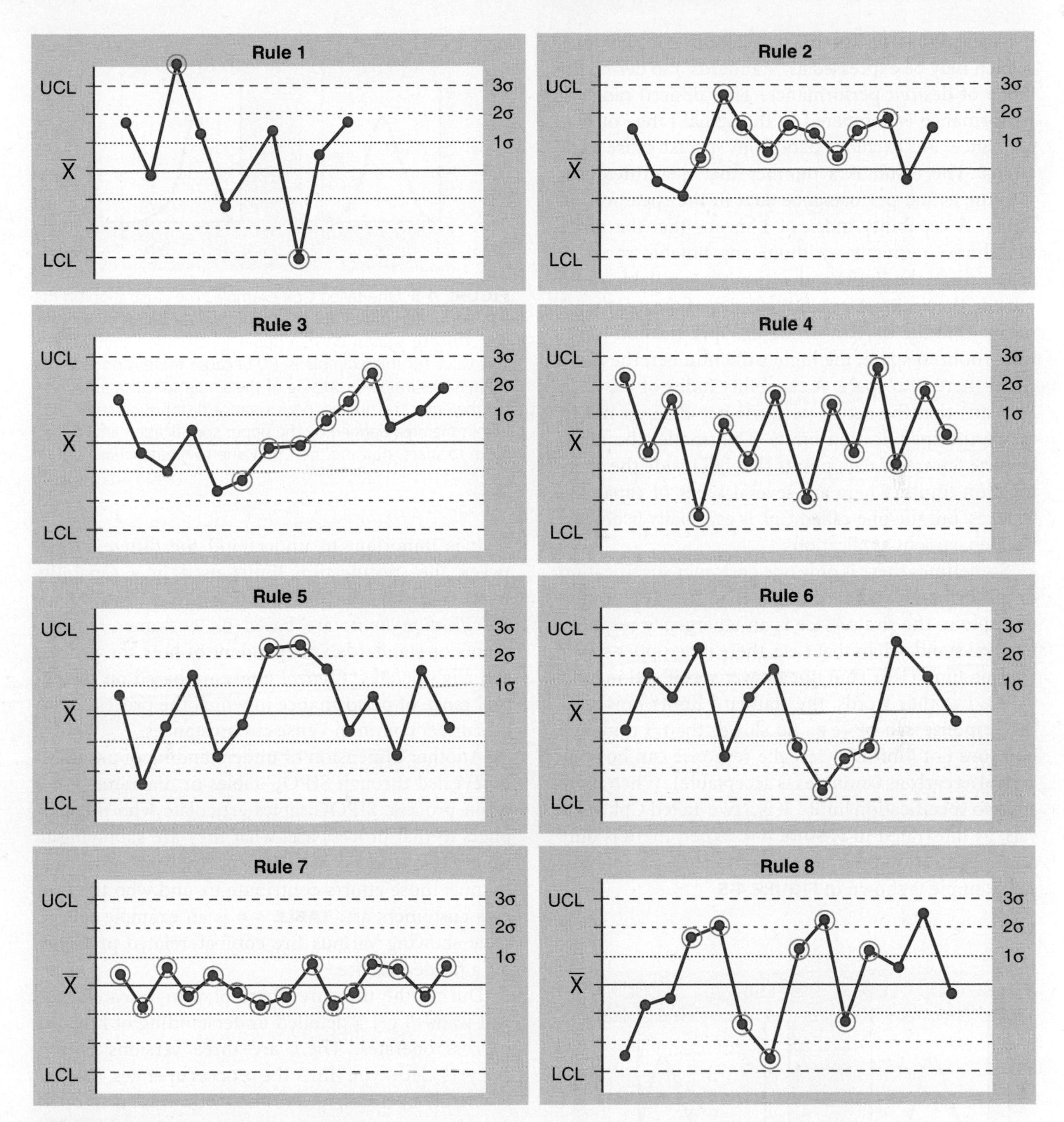

FIGURE 4-3 Special-cause pattern criteria.

chart for 1 month of turnout-time data. **FIGURE 4-3** shows a list of data patterns that signal the presence of special-cause variation.

Note that there are two places in Figure 4-3 where special-cause variation seems to be taking place. There is a day on which the turnout times exceed the UCL. There is also a period in which the data are elevated for several days but do not exceed the control limits, but the data do meet one of the criteria for special-cause data patterns.

To get a sound understanding of the performance level for a process, you need to look at the data without the special causes included. That level of performance gives you a baseline that you can then contrast with the requirements of the internal or external customers of the process or a performance standard that expresses those requirements. There is a very useful tool that quantifies how well the performance of the process met the process standards over a specified period of time. It is called a **capability index**.

A capability index uses customer specifications (which may be expressed as "standards") to define the range of desired performance. That desired range of performance is compared to the actual range of performance and excludes data from special-cause variations. The result is a number that quantifies how well the actual performance data fit the specification limits. A capability index of 1 means that the range fits almost perfectly—without any notable room to spare inside the limits and without any notable room outside the limits. A capability index of greater than 1 means that the data fit inside the specification limits with room to spare; the higher the number, the more room there is to spare. A capability index of less than 1 means that the actual performance data do not fit inside the specification limits. The smaller the number, the fewer the data points that fall inside the specification limits. There are several types of capability indexes, but the one called *Cpk* is especially useful for fire department applications.

Sometimes there is only one customer specification or performance standard rather than two. A common example in the fire service is an emergency response interval standard. In that case, there is an upper specification limit (USL) but not a lower specification limit (LSL). In other words, the standard limits how slow the response can be (e.g., no slower than 5 minutes) but does not limit how fast the response can be (e.g., arrival as early as 0 minutes is acceptable). When there are two specification limits, it is a two-tailed Cpk analysis, as illustrated in **FIGURE 4-4**. When there is only one specification limit, it is a one-tailed Cpk analysis. An example is shown in **FIGURE 4-5**.

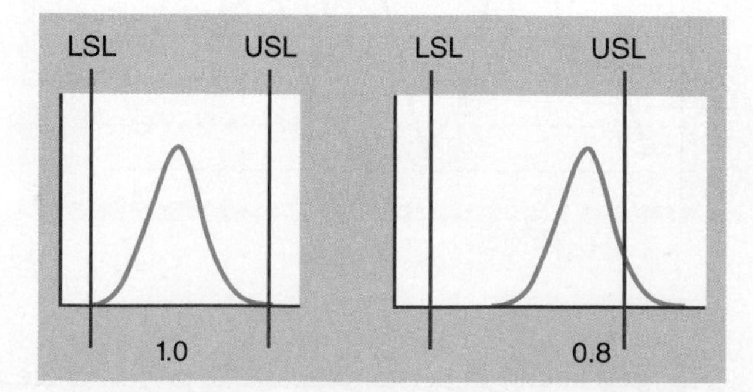

FIGURE 4-4 Two-tailed capability index (Cpk) example. The illustration on the left shows a distribution curve depicting the range of process performance against the upper and lower specification limits. The Cpk value for this example is 1.0 because the distribution fits inside the specification limits and there is no room to spare. On the right, some of the process performance distribution curve falls outside of the specification limits, pushing the Cpk value to less than 1.

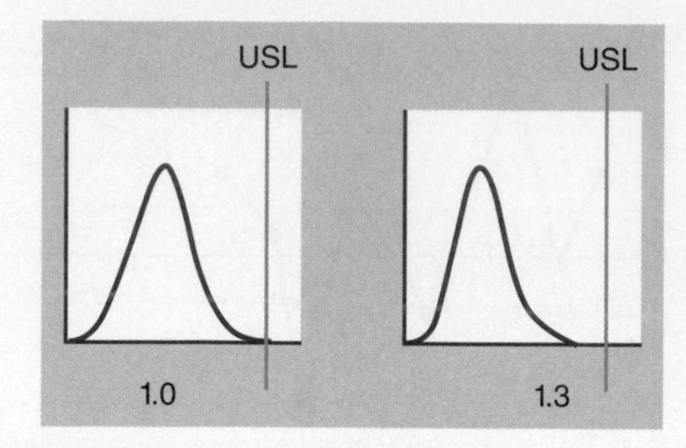

FIGURE 4-5 One-tailed Cpk example. The illustration on the left shows a distribution curve depicting the range of process performance against only an upper specification limit. The Cpk value for this example is 1.0 because there is no room to spare either inside or outside of the single specification limit. On the right, the process performance distribution curve fits within the area defined by the upper specification limit with room to spare, pushing the Cpk value to greater than 1.

It is important to understand the difference between the specification limits used in a capability index and the control limits used in an SPC chart. *Specification limits* are based on customer requirements or standards, independent of how the process actually operates. *Control limits* are based on the actual range of performance in which the process operates under common-cause circumstances.

Another dimension of understanding of processes is revealed through SIPOC tables or diagrams. For a given process, SIPOC tables articulate who the suppliers to that process are, what they are supplying to the process (inputs), what process they perform, what outputs those efforts contribute to, and who the process customers are. **TABLE 4-4** is an example SIPOC table showing various fire combat–related processes for a residential fire.

During the measure phase of your research, you will want to get a detailed understanding of how the process operates. There are three versions of this story: (1) How you think the process operates, (2) how it actually operates, and (3) how you want it to operate. How you want the process to operate will be the focus of your efforts later, during the improve phase of the project. How you think the process operates is prone to assumptions that may or may not be true. So your focus at this point should be on an objective examination of how the process *actually* operates. To document that, you will use flowcharts. You have probably seen these used to document a process for troubleshooting equipment problems or to outline the recommended process for evaluating and treating cardiac arrest victims (**FIGURE 4-6**). The key to making

TABLE 4-4 SIPOC Table for a Residential Fire-Attack Process

S (Supplier)	I (Input)	P (Process)	O (Output)	C (Customer)
911 Dispatcher	Information from caller	Deployment	Call location and incident data	Responding crews
Hydrant person	Hydrant, tools, supply line	Water supply to pump	Water supply	Pump operator
Pump operator	Pump settings	Water supply to hose	Charged attack line	Fire-attack crew
First-entry attack team	Charged attack line	Fire attack	Extinguishment	Property owner
	Access to structure or scene	Primary and secondary searches	Find and rescue remaining occupants	Remaining occupants

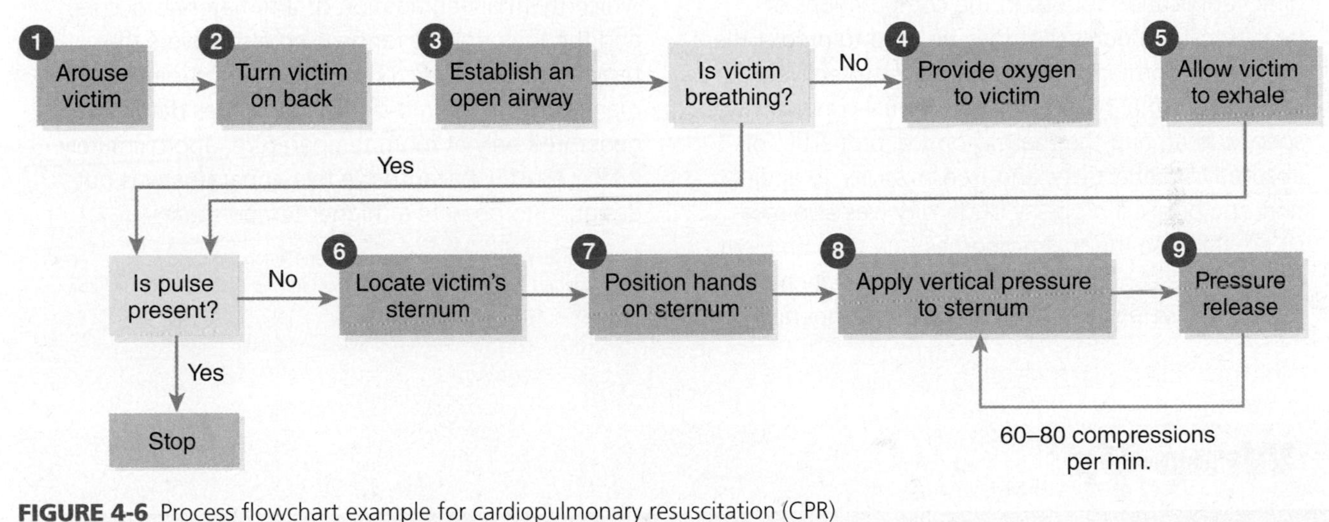

FIGURE 4-6 Process flowchart example for cardiopulmonary resuscitation (CPR)

Data from Florida State University Learning Systems Institute. Accessed April 12, 2009, www.lpg.fsu.edu/charting/InstructionalStrategies/howto-tactics/ht-sraflow.asp.

an accurate process flowchart is to observe the process directly and map what really happens—not what you think happens or what you want to happen. This "reality" version of process flow will be used later on when you start thinking about where there may be better ways to get things done as the basis for changes that could be implemented during the improve phase.

Another powerful tool to have in your arsenal during the measure phase is a *value stream map* of your process. This is a variation of a basic flowchart that includes information on inputs, outputs, and how much time is spent actually working on something versus waiting for the next step to occur.

For example, hiring a new person to fill a position vacated by an unexpected resignation can take a lot of time. There are often many delays in the hiring process, and those delays can be very expensive in terms of overtime costs, as well as burdensome to existing personnel who may not want to work the overtime hours. A high-level value-stream map shows the steps in the process and how much of the process is time wasted while waiting for the next step (**FIGURE 4-7**).

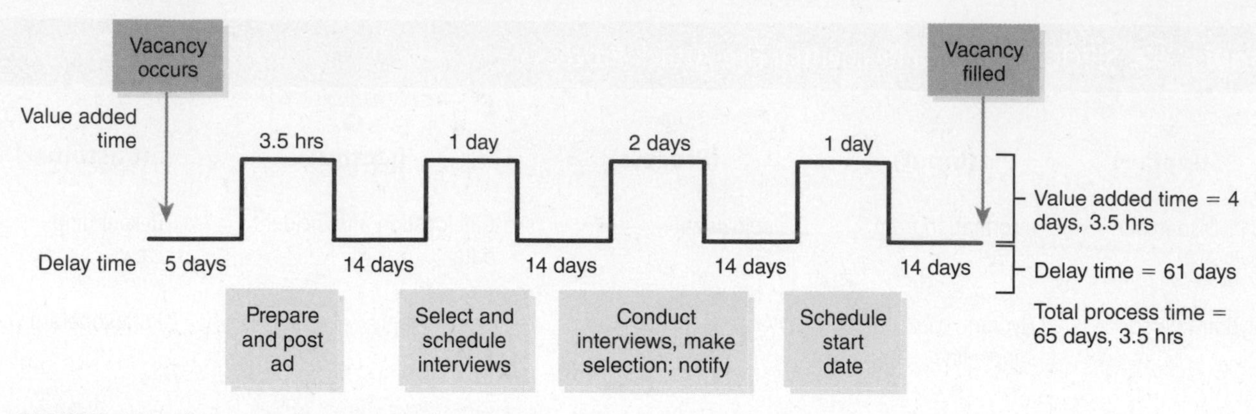

FIGURE 4-7 Value-stream map for a hiring process.

THE RESEARCH EXPERIENCE

A series of measurements has been carried out to quantify the thermal properties of materials used to fabricate fire fighters' thermal protective clothing. The thermal property measurements chosen have direct application for use in the computations of heat-transfer models that may be used to predict the thermal performance of fire fighters' protective clothing. The thermal properties are thermal conductivity; specific heat; and the thermo-optical properties of absorptivity, reflectivity, and transmissivity. In addition, the physical property of density was also measured. Because thermal properties vary with ambient temperature conditions, thermal conductivity measurements were made over a range of temperatures, 20°C to 100°C (68°F to 212°F). Specific heat was measured over a range of 0°C to 100°C (32°F to 212°F). The maximum temperature of 100°C (212°F) was selected because it is below the temperature at which thermal degradation of the materials occurs, and the temperature range used also covers the temperatures at which burn injuries occur in human skin. The thermo-optical properties data were generated only at room temperature, approximately 23°C (73.4°F), because the test apparatus was not designed to operate at higher temperatures.

Lawson, J. R., Walton, W. D., Bryner, N. P., and Amon, F. K., *Estimates of Thermal Properties for Fire Fighters' Protective Clothing Materials*, NISTIR 7282, June 2005.

Wrap-Up

CHAPTER SUMMARY

This chapter has provided the tools necessary to be able to define the problem. If you are not able to define the problem, you will not be able to solve the problem. Once you have defined the problem, you need to be able to measure the problem. A few specific models were discussed in this chapter, along with processes designed to allow you to measure the problem. If the problem is not measurable, it is not a problem that can be solved with a sound answer.

KEY TERMS

Baseline How a process currently operates.

Capability index A useful tool that quantifies how well the performance of the process met the process standards over a specified period of time.

Define phase Clarifies the problem or issue you are trying to address.

Fire Service Performance Indicator Format (FSPIF) model Addresses the issue of standardized formatting for performance indicators and selecting an indicator format.

Improve phase Uses model programs to evaluate the effectiveness of the program to identify areas to improve.

Measure phase A detailed assessment of the current status of the process (or processes).

Performance indicator Shows how well (quality) or how efficiently (economically) a process is performing.

Return on investment Projects that have the most important impact for the customer or that have the greatest potential to provide a significant impact on the organization in relation to the money and time invested. ROI may be operational, financial, political, or clinical.

REVIEW QUESTIONS

1. Identify a problem that, if you could solve it, would provide ROI for your fire department. Once you have identified the problem, write a problem statement for it.
2. Describe the components of the FSPIF. Explain each component.
3. Explain the SIPOC model and how it is used in the measurement process.
4. During the improve phase, you can use a flowchart to illustrate how the process actually works. Using a procedure that is currently in place in your fire department, develop a flow chart to illustrate how the process works.
5. Describe the capability index.
6. Describe how you define a problem as it relates to research.
7. Discuss the various ways to measure a problem.
8. Identify various models used in measuring.
9. What is a performance indicator?
10. Describe ROI and its importance in relation to your research project.

REFERENCES

Flynn, J. 2009. *Fire Service Performance Measures.* Quincy, MA: National Fire Protection Association. Accessed: June 23, 2020, http://tkolb.net/FireReports/FireServicePerformanceMeasures2008.pdf.

Gunderson, M. 2009. "Performance Indicators." In *Evaluating and Improving Quality in EMS,* edited by E. B. Lerner, R. C. Pirallo, R. A. Swor, and L. J. White, 99–113. Dubuque, IA: Kendall/Hunt.

iSixSigma. n.d. "Voice of the Customer (VOC)." Accessed June 15, 2020, http://www.isixsigma.com/dictionary/Voice_Of_the_Customer_(VOC)-391.htm.

National Association of State EMS Officials and National Association of EMS Physicians. 2006. "EMS Performance Measures Project: Recommended Attributes and Indicators for System/Service Performance."

The Joint Commission. "Important Modifications—Performance Measurement Reporting." n.d. Accessed June 15, 2020, https://www.jointcommission.org/measurement/.

Wikipedia. "Voice of the Customer." 2020. Accessed June 15, 2020, http://en.wikipedia.org/wiki/Voice_of_the_customer.

CHAPTER

Analyzing Baseline Results

LEARNING OBJECTIVES

Upon completion of this chapter, you should be able to:

- Explain the rationale and process used to analyze baseline results.
- Describe the various tools used to analyze baseline results.

Case Study

The Somewhereville Fire Department decided to start routinely measuring the performance of its fire ground operations with several key performance indicators. These will provide a baseline so that as various improvement projects are undertaken in the future, the department can better assess the impact of these efforts over time. It chose the following key performance indicators on residential structure fires:

- Interval from time on-scene to first water on fire
- Interval from time on-scene to primary search completed
- Interval from time on-scene to secondary search completed

To capture these times, fire crews were instructed to announce them on the radio. Upon hearing these announcements, the dispatchers timestamp the events using preconfigured function keys for the computer-aided dispatch (CAD) system. At the end of each month, the times are harvested from the CAD as a file, which is then added into a spreadsheet file with other results. Formulas in the spreadsheet calculate the time intervals on each call. Periodically, each crew gets a performance feedback report showing their time intervals on each residential fire call. The time intervals on each fire are also plotted on a run chart so that crews can see if their performance is improving, declining, or staying about the same. The crew performance is plotted beside other lines on the run chart showing the average time intervals for the other battalions and the entire department. Battalion chiefs and command staff see results aggregated by battalion and the overall department.

1. What key performance indicators, if any, does your department routinely measure? Are any of these performance measures specifically for structure fires?
2. What other indicators, besides the ones described in the Case Study, would you suggest be measured on all structure fire calls to help crews get feedback on their fire ground performance?
3. Do you see any downside to tracking performance in fire ground operations?
4. What resistance, if any, would you anticipate from crews or command staff in tracking key performance indicators of fire ground operations on structure fires?

Introduction

The *define* phase of the project outlines the problem. The *measure* phase provides an understanding of how the process works and what its current level of performance is. In the *analyze phase,* you will try to determine the causes(s) of problems or what factors may be holding the process back from reaching higher levels of performance.

Analyze Phase

The tools used in the **analyze phase** of a project include the following:

- Cause-and-effect diagrams
- Brainstorming
- Pareto diagrams
- Histograms
- Correlation analysis
- Regression analysis

Cause-and-Effect Diagrams

One of the most useful tools for identifying potential causes of problems or hindered performance is the **cause-and-effect diagram** (American Society for Quality 2020). It is also called a *fishbone* or *Ishikawa diagram.* Identifying the potential causes will help you change or modify the relevant parts of the process to improve performance.

Suppose your problem statement describes a growing problem with long response times for emergency calls. To help you think through what all of the possible causes may be, list several broad categories that apply to the process you are addressing. You may also try to use one of the generic problem-category lists shown in **TABLE 5-1**.

To create a cause-and-effect diagram, begin by drawing a horizontal line across the middle of a sheet of paper, whiteboard, flip chart, or computer screen. Software tools for diagramming business processes are available that support building cause-and-effect diagrams—for example, QI Macros for Excel (2020) and Microsoft (n.d.) Visio—but paper works well too. The line should end with an arrowhead pointing to the problem you are addressing. This is the spine of the diagram. Using one of the category sets from Table 5-1, or any other set of categories that makes sense to you, make each one of those categories a separate line branching off at a vertical slant from above or below the horizontal line. These slanted lines are the ribs of the diagram. The potential causes are then labeled on lines coming off the ribs. You should end up with a diagram similar to the one shown in **FIGURE 5-1**.

Brainstorming

Brainstorming (Wikipedia 2020) is a useful technique for coming up with ideas of potential causes or limiting factors to include in the cause-and-effect diagram. To make brainstorming sessions as productive as possible, several general rules are often followed to help generate ideas, encourage creativity, and make everyone feel comfortable participating:

- The more ideas, the better. As more and more ideas are discussed, even seemingly silly ones, the more likely it is that a useful or relevant idea will be suggested.
- Filter later. To maximize the number of ideas (per the previous point), don't give people reasons to filter their ideas before expressing them. Try to get everyone to let ideas flow freely and unfiltered. The ideas can be sorted for usefulness later.
- Be creative. Try to look at the problem or challenge from different perspectives. How would a different company or another industry deal with this issue? How would the characters from your favorite TV show, movie, or cartoon tackle it? New ways of looking at the issue can stimulate new ideas.

TABLE 5-1 Generic Problem-Category Lists for Use in Cause-and-Effect Diagrams

■ What	■ Manpower	■ People	■ Equipment
■ Why	■ Materials	■ Policies	■ Process
■ Where	■ Methods	■ Procedures	■ People
■ When	■ Machines	■ Place	■ Materials
	■ Measurements		■ Environment
	■ Environment		■ Management

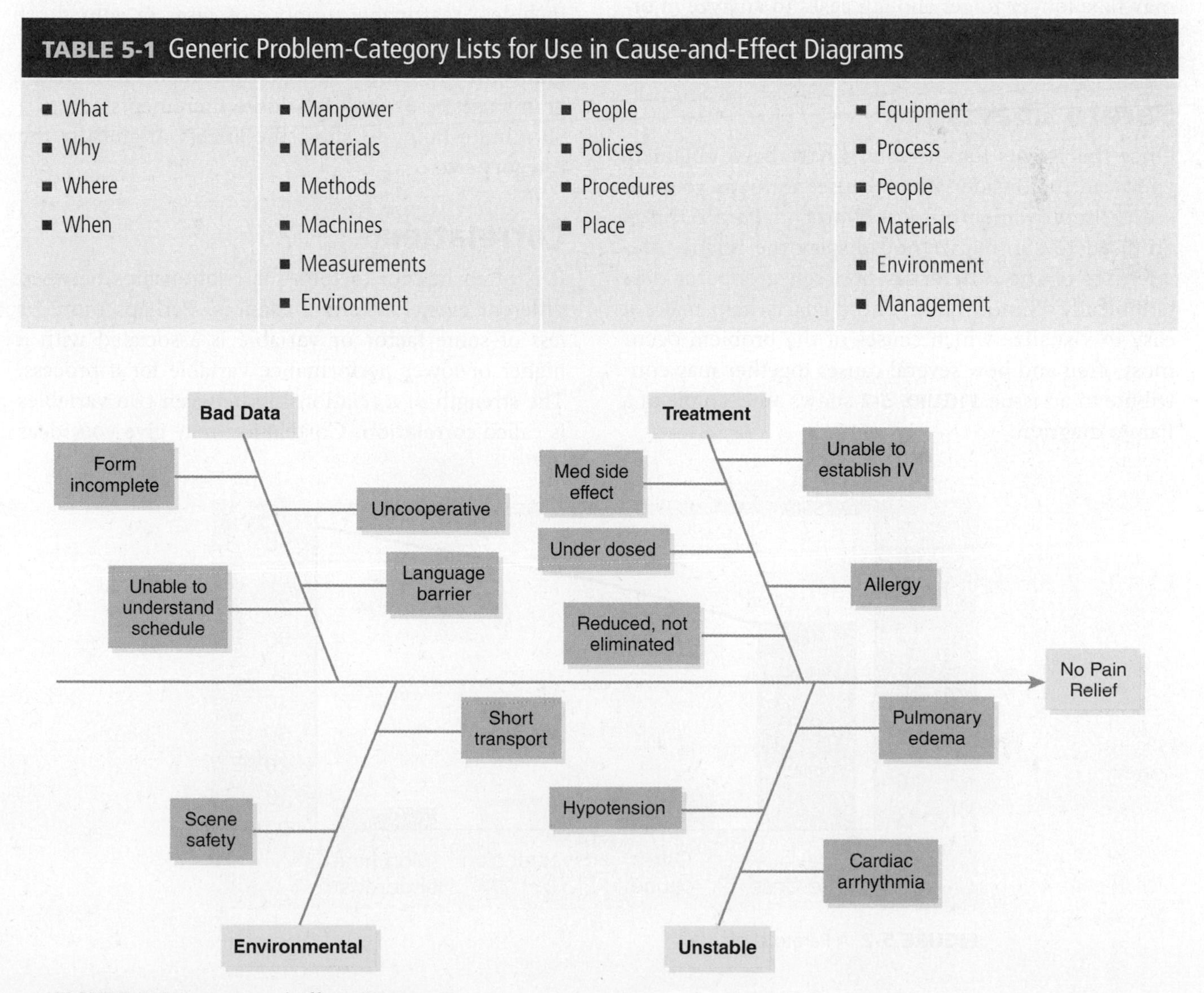

FIGURE 5-1 A cause-and-effect diagram.

- Build on ideas or consolidate them. Look at the ideas that have been mentioned so far. One may trigger an idea for something that expands or builds on it. Look at similar ideas: A new one may be a combination of two or more that have already been mentioned.

Now that we have a list of the potential causes from our brainstorming session captured on a cause-and-effect diagram, the challenge shifts to counting how often each of the items on the cause-and-effect diagram actually occurs. Retrospectively, data from calls or situations in the past where the problem occurred could be analyzed to count how often each of the various causes occurred. If retrospective data are not available or the level of detail needed is absent, the counts can be made prospectively for a period of time. For a frequently occurring problem, such as run report errors, that period of time might be a week or so (depending on call volume). A less frequent problem event, such as employee resignations, may take longer to get enough cases to analyze in order to get a reasonable count of causes.

Pareto Charts

Once the counts for the causes have been obtained, you need to consider which causes to focus your research/improvement-project efforts on. **Pareto charts** are used to summarize and display the relative frequencies of the differences between groups of data graphically (Simon n.d.). These charts help make it easy to visualize which causes of the problem occur most often and how several causes together may contribute to an issue. **FIGURE 5-2** shows an example of a Pareto diagram.

Histograms

Another useful way to help analyze data visually is with a **histogram**. A histogram shows the proportion of cases that fall into adjacent, nonoverlapping categories. The number of cases that fall into each category is shown on a bar. The more cases fall into a category, the taller will be the corresponding bar. Histograms allow you to see how the data are distributed across the categories. This can give you additional insights into where the problems exist or where limitations are hindering your process. The size of the category can be varied to suit the level of detail desired.

Response-time performance data are often shown in histograms. The categories are increments of time. Of course, creating a histogram, say, for each second across a range of 30 minutes for response times would require 1800 separate bars—which is far more detailed than will be useful. However, a histogram with 15-minute increments is too coarse—it would have only two bars. Pick a category size that lets you include a reasonable number of cases in your most frequent categories. **FIGURE 5-3** shows several histograms for the same response-time data. These histograms use 1-, 5-, and 8-minute increments across a 30-minute range to show the effects of altering the category size.

Correlation

It is often helpful to look for relationships between different events or circumstances. Perhaps more or less of some factor or variable is associated with a higher or lower performance variable for a process. The strength of a relationship between two variables is called **correlation**. Correlation may give you ideas

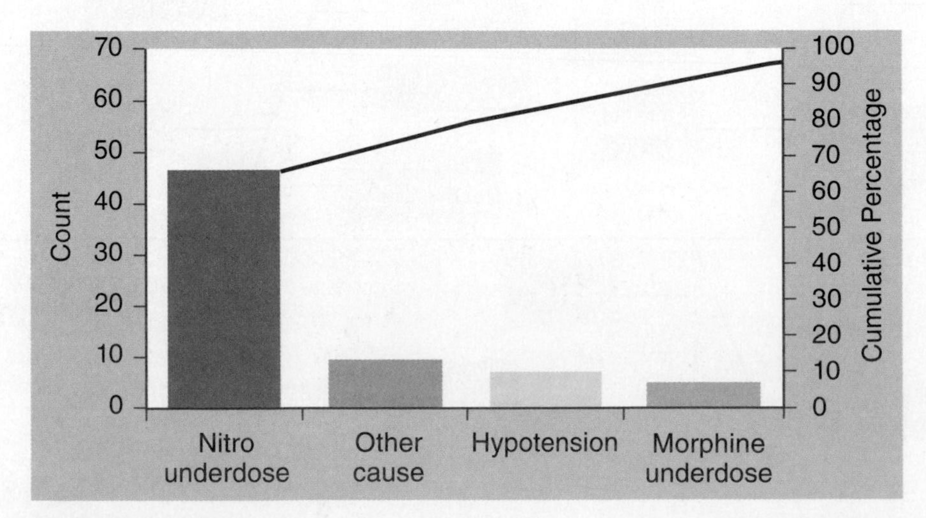

FIGURE 5-2 A Pareto chart.

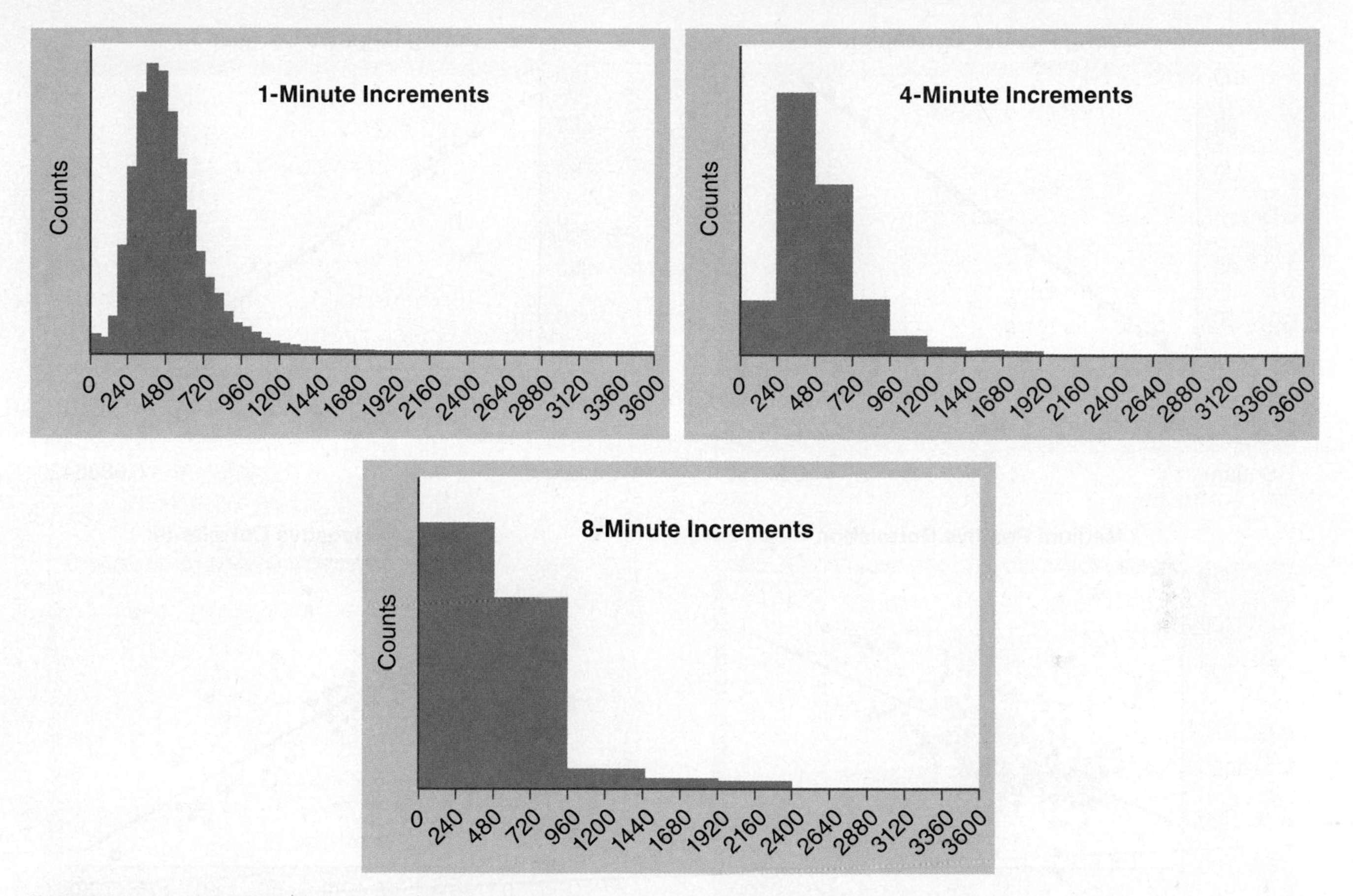

FIGURE 5-3 Histograms in 1-, 5-, and 8-minute increments over a 30-minute range.

about things that you might try to control or influence to get better performance.

Correlation is expressed as a value called a **correlation coefficient**. This value will range anywhere between or including 1 and −1. The closer the correlation coefficient is to 1 or −1, the stronger is the relationship. The closer the correlation is to 0, the weaker is the relationship. In a *positive correlation*, as the value of one of the variables increases, the value of the other also increases. Positive correlations have correlation coefficients greater than 0. In a *negative correlation*, as one of the values increases, the value of the other variable decreases. Negative correlations have correlation coefficients that are less than 0. Correlation can be displayed visually by plotting the two variables, as shown in **FIGURE 5-4**.

Consider a project in which you are trying to reduce employee turnover. You might want to look at traits that can be correlated to employee longevity. You would use length of employment as one variable and then calculate the correlation coefficient between it and other variables at the time the applicant is being considered. Those other variables might include the following:

- Number of college credits completed
- Hiring examination scores
- Years of any type of prior employment experience
- Years of prior emergency services experience
- Height
- Physical agility test scores

If you saw a strong positive correlation between the number of college credits completed and how long people stay working for your organization, you might want to explore whether that should be a factor to consider in your employee hiring process to potentially reduce turnover. However, there is a very important caveat to remember about correlation: It does not necessarily mean that there is a cause-and-effect relationship; that is, it does not mean that changing one variable will cause the other variable to change.

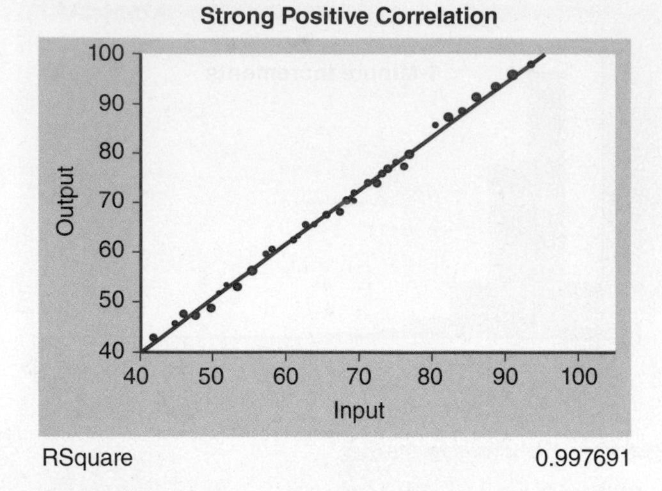

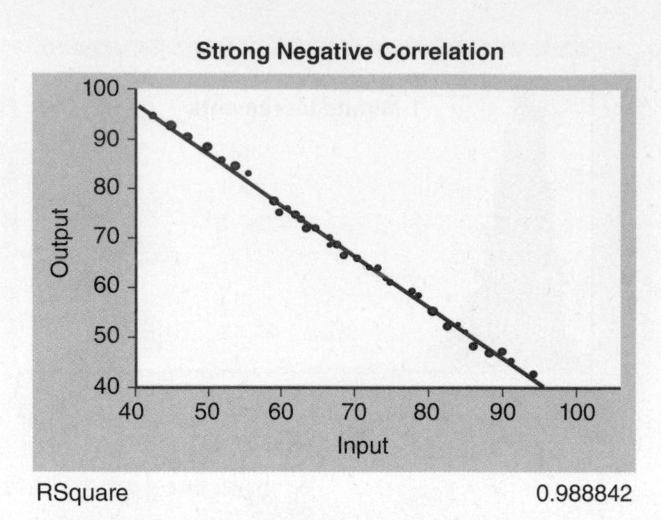

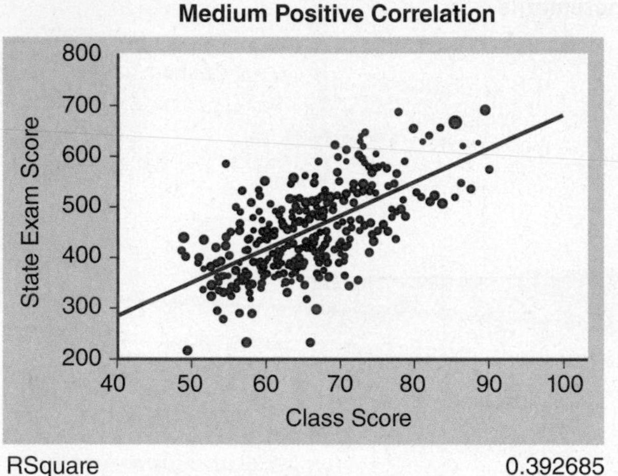

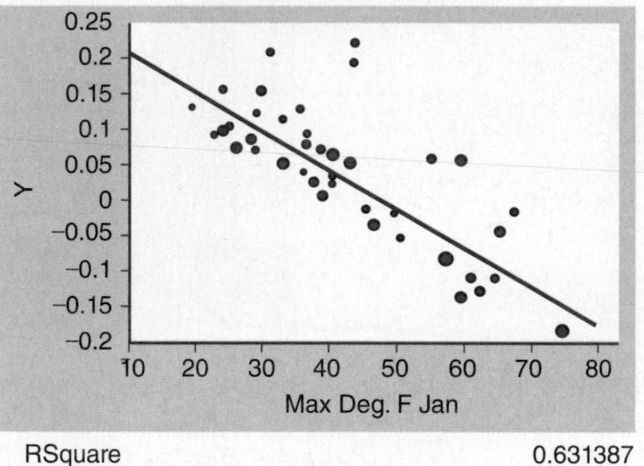

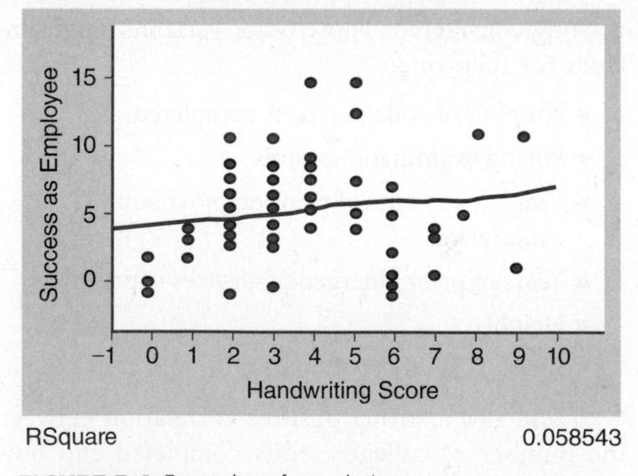

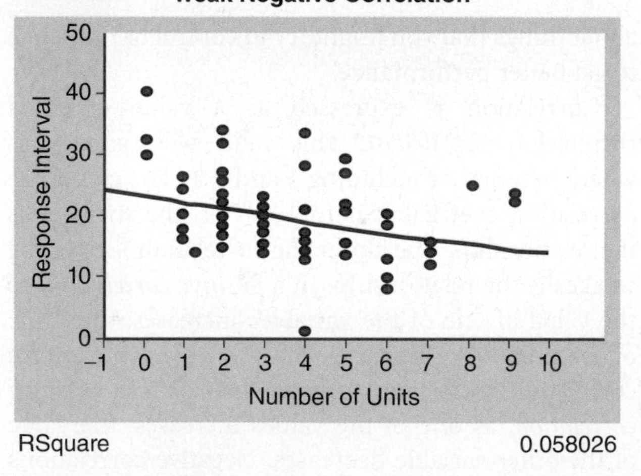

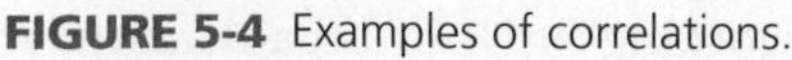
FIGURE 5-4 Examples of correlations.

Correlations may be coincidental. You might find a strong positive correlation between height and employment longevity. That does not necessarily mean, however, that taller people will want to spend more of their careers working for your organization. You might find a strong negative correlation between the amount of ice cream eaten and the murder rate, but that does not mean that eating less ice cream will prevent murders. It also does not mean that increasing ice cream sales will be an effective method for reducing the murder rate.

It may also be possible that both of the variables are affected by a third variable. The hardness of a block of ice cream left on a picnic table for 15 minutes may be negatively correlated to the rate of growth of grass in your yard, but that does not mean that having soft ice cream on the picnic table will make the grass grow faster. It might be that higher outdoor temperatures, the third variable, cause both the ice cream to get soft and make your grass grow faster. Correlation simply shows where there are relationships that may warrant further exploration for cause-and-effect linkages.

Regression

When your consideration of how a process operates suggests that there may be a mechanism or rationale for a cause-and-effect relationship between an input variable that you can control or influence and that results in an output variable, you can move on to regression analysis.

Consider the problem of survival after cardiac arrest. Our understanding of the processes at work in cardiac arrest and resuscitation efforts tell us that earlier interventions will improve the chances for survival. Correlation analysis shows a good negative correlation—faster response times correlate with higher survival rates. This is the right type of circumstance to have to apply regression analysis. Suppose your community has a goal of achieving a 30% rate of survival to hospital discharge in cases when someone witnessed a patient collapse and the patient was found by the first-arriving crew to be in ventricular fibrillation. If response times (from the estimated time of collapse, calculated as 1 minute before the time that 911 was called, to the time that the initial defibrillation was delivered) were analyzed for correlation with survival rates, a correlation coefficient of 0.8 might be found. However, this −0.8 correlation coefficient does not tell you what response time will be needed to achieve the target of a 30% survival rate. **Regression analysis** is a tool that allows predictions of outcomes to be made on the basis of changes to an input variable. The result of a regression analysis is a regression equation, which can be shown graphically as a regression line.

FIGURE 5-5 shows a regression line calculated using data from the Seattle and Tucson fire departments. Assuming that public education efforts with telephone prearrival instructions can get bystander cardiopulmonary resuscitation (CPR) under way in 1 minute after a person collapses, the input variable, shown along the horizontal axis (x-axis), is the estimated collapse-to-defibrillation time. The output variable, shown on the vertical axis (y-axis), is the survival rate. The regression line shows that a collapse-to-defibrillation interval of somewhere between 6 and 7 minutes will be needed.

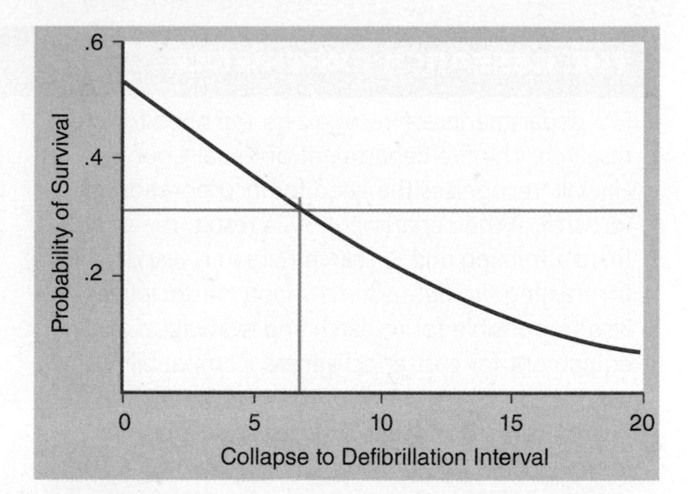

FIGURE 5-5 A regression line calculated using data from the Seattle and Tucson fire departments. Assuming that public education efforts with telephone prearrival instructions can get bystander cardiopulmonary resuscitation (CPR) under way in 1 minute after a person collapses, the input variable, shown along the horizontal axis (x-axis), is the estimated time from collapse to defibrillation. The output variable, shown on the vertical axis (y-axis), is the survival rate. The regression line shows that a collapse-to-defibrillation interval of somewhere between 6 and 7 minutes will be needed.

Similar correlation and regression analyses could be made between input variables and output variables, such as response-time intervals and dollar loss on structure fires or the time interval since the last fire inspection at a commercial property and the probability of that property having a fire.

Other Types of Statistical Analyses

As your experience, skill, and insight develop, you may be able to apply a much broader set of statistical tools to problem analysis. This text cannot provide an entire course in statistics, but textbooks on statistics, from beginner to advanced levels, can be purchased through any campus or retail bookstore. You can also explore the wide variety of textbooks on statistics specific to research and process improvement. Many websites also offer excellent resources for learning about statistics. The Additional Resources section at the end of this chapter lists some of the resources that fire service professionals who are getting started in research and process improvement may find helpful.

The insights you gain from the analysis phase will provide ideas to use in the next phase: *improvement.*

THE RESEARCH EXPERIENCE

Fire departments are recognizing the need for more research. The fire department of Kauai County, Hawaii, recognizes the need for incorporation of research in the department. As a result, the department's Training and Research Bureau is responsible for training fire fighters in a variety of disciplines. It is also responsible for research and evaluation of new equipment for cost-effectiveness, compatibility, and safety.

The objectives of the Training and Research Bureau include developing and maintaining a proficient, professional, and structured training program that will enhance services to the community and safety for department employees. The bureau utilizes research in order to meet the following objectives:

1. Review and update department training manuals.
2. Standardize drills to give personnel structured training throughout the department.
3. Enhance safety programs. Train fire fighters to think and practice safety at all times on the job.
4. Utilize computer programs to maintain complete and accurate drill records for all personnel.
5. Integrate water safety officers and their training into the department.
6. Adhere to national standards of incident command training and implementation.
7. Train to the highest medical standards possible.

Kauai Fire Department, Training and Research Bureau. Accessed January 8, 2009, www.kauai.gov/Government/Departments/FireDepartment/TrainingandResearchBureau/tabid/125/Default.aspx.

Wrap-Up

CHAPTER SUMMARY

This chapter took us through the second step, the analyze phase, by looking at a variety of ways to analyze a problem. The tools used in the analyze phase of a project included cause-and-effect diagrams, brainstorming, Pareto diagrams, histograms, correlation analysis, and regression analysis. These tools give us the ability to look at a problem in such a way as to gain a better understanding of the issues at hand.

KEY TERMS

Analyze phase Determining the causes(s) of problems or what factors may be holding the process back from reaching higher levels of performance.

Brainstorming A useful technique for coming up with ideas of potential causes or limiting factors to include in a cause-and-effect diagram.

Cause-and-effect diagram A tool for identifying potential causes of problems or hindered performance. It is also called a *fishbone* or *Ishikawa diagram*.

Correlation The strength of a relationship between two variables.

Correlation coefficient Correlation expressed as a value between −1 and 1.

Histogram Shows the proportion of cases that fall into adjacent, nonoverlapping categories.

Pareto chart Summarizes and displays graphically the relative importance of the differences between groups of data.

Regression analysis A tool that allows predictions of outcomes to be made on the basis of changes to an input variable.

REVIEW QUESTIONS

Use the following list of analysis tools to answer Questions 1–4:

- Cause-and-effect diagrams
- Brainstorming
- Pareto diagrams
- Histograms
- Correlation analysis
- Regression analysis

1. Describe each of the tools.
2. Explain how to use each of the tools.
3. Which tool would be best to use in a research project to illustrate the number of fire deaths of elderly patients in nursing homes? Why?
4. Which tool would be best to use in a research project to illustrate response times? Why?
5. Explain regression lines.
6. Explain the rationale and process used to analyze baseline results.
7. Explain the purpose of the analyze phase.
8. What numerical value constitutes the weakest relationship in a correlation coefficient?
9. Describe a fishbone or Ishikawa diagram.
10. What variables would you consider if you were doing a research project on response times?

REFERENCES

American Society for Quality. 2020. "Fishbone Diagram." Accessed June 15, 2020, http://www.asq.org/learn-about-quality/cause-analysis-tools/overview/fishbone.html.

Microsoft. n.d. "Visio." Accessed June 15, 2020. https://www.microsoft.com/en-us/microsoft-365/visio/flowchart-software.

QI Macros for Excel. 2020. "Fishbone Diagram Template for Excel." Accessed June 15, 2020, https://www.qimacros.com/fishbone-diagram-template/.

Simon, K. n.d. "Pareto Chart." iSixSigma. Accessed June 15, 2020, www.isixsigma.com/library/content/c010527a.asp.

Wikipedia. 2020. "Brainstorming." Accessed June 15, 2020, http://en.wikipedia.org/wiki/Brainstorming.

ADDITIONAL RESOURCES

Blakstad, Oskar. 2008. "Statistics Tutorial" (February 13). Explorable.com. Accessed June 24, 2020, https://explorable.com/statistics-tutorial.

Calkins, Keith G., Shirleen Luttrell, assist. 2010. *An Introduction to Statistics* (Fall). Accessed May 20, 2020, http://www.andrews.edu/~calkins/math/webtexts/statall.pdf.

Rumsey, Deborah J. 2005. *Statistics Workbook for Dummies.* Hoboken, NJ: John Wiley & Sons, Inc.

Rumsey, Deborah J. 2009. *Statistics II for Dummies.* Hoboken, NJ: John Wiley & Sons, Inc.

Rumsey, Deborah J. 2011. *Statistics for Dummies*, 2nd ed. Hoboken, NJ: Wiley Publishing, Inc.

Schmuller, Joseph. 2016. *Statistical Analysis with Excel for Dummies*, 4th ed. Hoboken, NJ: John Wiley & Sons, Inc.

CHAPTER

Improving the Process

LEARNING OBJECTIVES

Upon completion of this chapter, you should be able to:

- Describe the improvement process.
- Explain the difference between independent and dependent variables.
- Discuss ways to prevent bias in a study.
- Explain statistical power.
- Discuss the importance of the P value.

Case Study

The Somewhereville Fire Department adopted a team-based quality program. Each targeted area has a permanent improvement team to help track performance in that area, generate ideas for improvement projects, and manage the area's ad hoc improvement project teams. Several of these target-area teams were established, including the following:

- Fire attack operations
- Pump operations
- Fire prevention
- Fire investigation
- Cardiac arrest
- Stroke
- Data and analysis

1. If your department appointed you to the team handling fire attack operations, what hypothesis might you suggest to the team to improve your department's fire attack operations?
2. Answer the previous question for the pump operations team.
3. Answer Question 1 for any of the other teams listed in the Case Study.

Introduction

The *define* phase of the project outlines the problem. The *measure* phase provides an understanding of how the process works and what its current baseline level of performance is. The *analyze* phase provides insights into what factors may have been causing errors or holding the process back from reaching higher levels of performance. In the **improve phase**, you will choose which of these factors to change in an effort to improve performance.

Choosing What to Change

One of the first challenges in the improve phase is deciding which factors to change in hopes of improving performance. Some of the issues to consider for each factor include the following:

- How likely is it that changing that factor will improve performance?
- How much impact on performance can you expect to see if that factor is changed?
- How difficult will it be to change that factor?
- Do you have, or have access to, the expertise needed to change that factor?
- How much will it cost to change that factor?
- Are there other costs connected to changing that factor (e.g., training and monitoring the new process)?
- Do you have the resources, financial and otherwise, to change that factor?
- How much resistance will there be to changing that factor? From employees? From external customers? From the general public, media, and elected officials?
- How long will it take to make change(s) to that factor?
- Are there any potentially undesired collateral effects that changes to that factor may provoke?
- Can changes to that factor be sustained?

The factors to consider changing first are the ones that look best after considering the issues in the previous list. To minimize the risks, reduce costs, and make the change process easier to deal with, small-scale tests or pilot studies can be very helpful. If the pilot study shows good results, it can be expanded. If the pilot results are poor, the unfavorable consequences have been minimized, and something else can be attempted.

In research terminology, the factors that influence process performance are called **independent variables**. The measure of performance (i.e., the performance indicator) that you are trying to change is called the **dependent variable**.

In the terminology of performance improvement, each of the independent variables is referred to as an X. If there are several, they are numbered X_1, X_2, X_3, and so forth. The dependent variable is called a Y.

This stage of research (and the performance-improvement process) is about finding the right changes to make in one or more of the independent variables (or Xs) to get the desired change in the dependent variable (or *Y*). The terms *independent variable* and *X* can be used interchangeably. So can the terms *dependent variable* and *Y*.

Changes Are a Hypothesis

The change we make to an independent variable is a **hypothesis**. Our process change is a type of experiment. We are making the change because we think it will improve performance. The resulting change in performance, if it occurs, should manifest as a **statistically significant** change in the dependent variable. A statistically significant difference is one for which statistical tools show that the differences observed between performance before the change and performance after the change (or between one group and another group) are unlikely to be the results of chance—that the differences are real.

There is another type of variable we also need to deal with: **extraneous variables**. These are factors, other than the independent variables, that may influence the results we see in the dependent variable. When we test our hypothesis, we want to eliminate or minimize the influence of any extraneous variables.

Experimental Design

Let's consider an example to help illustrate these concepts and terminology. Suppose we are trying to improve the performance of the fire combat process at residential structure fires. During the define phase of the project, we identified that one of the measures for how well the process is working is the amount of time it takes after arriving on the scene to get water on the fire. This *first-water-on-fire interval* (FWOFI) variable is the dependent variable we will focus on for this project. One of the many factors we identified that might affect the FWOFI is the configuration of our preconnect attack lines.

A debate came up among the members of the project team regarding the impact that the size of the hose would have. It was argued by some that a smaller, 1½-in. (38-mm) hose could be brought in faster and thereby reduce the FWOFI. Others argued that a 1¾-in. (44-mm) hose could be brought in just as fast. The question of flow capacity is a separate issue. However, it was agreed by the project team members that if there is no difference between the two sizes of hose on the preconnect lines, then the larger hose will be preferable in case a greater capacity for water flow is needed.

The experiment for this project will test the hypothesis that the 1½-in. (38-mm) hose configuration on a preconnect line is associated with a shorter FWOFI than using a 1¾-in. (44-mm) hose (**FIGURE 6-1**). The hose size is the independent variable (or *X*) that will be manipulated to see whether there is any difference in the FWOFI, the dependent variable (or *Y*).

There are several ways to design a test of this hypothesis. Two of the many options are the following:

- Option A: Put the 1¾-in. (44-mm) hose on half of your department's apparatus and the 1½-in. (38-mm) hose on the other half. Over the next several months (or years, depending on how many residential structure fires you actually have), gather FWOFI data from each fire, and then compare results between the two groups.
- Option B: Set up an evolution at a fire training facility that simulates the circumstances at a residential structure fire. Have a crew run the exercise using the 1¾-in. (44-mm) hose and then try the 1½-in. (38-mm) hose.

There will be several extraneous variables that might influence the results for Option A. The fires to which the crews using the 1¾-in. (44-mm) hose respond might be more serious than the others; fire severity is an extraneous variable. The fires may be deeper inside the structure and therefore take longer to reach; the location of the fire is another extraneous variable. There may be differences in the experience of the crews; crew experience level is another extraneous variable. There may be other factors that change over the year during which the project is conducted, such as new employees, new apparatus, or changes to other equipment or operating procedures. These are also extraneous variables. The effect that extraneous variables have on the dependent variable is called **bias**.

FIGURE 6-1 Conducting an experiment using a 1½-in. (38-mm) hose versus 1¾-in. (44-mm) hose.

By conducting the experiment at the fire training facility and in the same location inside the training building, many of the extraneous variables for Option A can be eliminated or minimized. There will be no change in the severity of the call. There will be no difference in the location of the fire. The evolutions can be conducted within a relatively short period of time, thereby reducing the potential for new employees, new apparatus, or procedure changes to bring bias into the results. However, some extraneous variables remain for Option B. The crews will be less fatigued on their first evolution using the 1¾-in. (44-mm) line, possibly making the first evolution go faster. The test sequence is an extraneous variable.

For many of the extraneous variables we have described, there are ways to design the tests so that the effect of the extraneous variable is nullified. For example, if the evolutions are repeated on a sufficient number of occasions, the order of which hose is used first could be alternated. An even better design would randomly designate which hose is used each time the evolution is conducted. The evolutions are conducted enough times to ensure that each hose is used first the same number of times. This is called **randomization**.

Some sources of bias are more subtle. Suppose we are trying to improve the extinguishing process for wood and paper fires. We might test a hypothesis that a dose of commercial wetting agent put into tank water works better than an equivalent dose of simple dishwashing liquid. The type of wetting agent is the independent variable. The time to flames extinguished is the dependent variable. However, what if some of the crews have been convinced by advertising that the commercial wetting agent is better, even though there isn't any objective research to support the manufacturer's claim? This may create the potential for those crews to fight the fire less aggressively, consciously or unconsciously, when using the dishwashing liquid—thereby introducing a bias into the results.

One way to prevent project participants from intentionally biasing the results is to prevent them from knowing which version of the process change they are using. This called **blinding**. In a *single-blind* experimental design, the users of the process do not know which version of the independent variable is being used at any particular time. In a *double-blind* experimental design, the participants as well as the people collecting the data do not know which version of the independent variable is being used.

The potential for bias can also be reduced by using a **placebo**. A placebo is a version of the independent variable that has no effect on the dependent variable. For example, instead of a commercial wetting agent or dishwashing liquid, a placebo of plain tap water could be added to the tank water.

The experiment can be designed so that on a random basis, 16 ounces (473 mL) of commercial wetting agent, dishwashing liquid (in appropriate dilution), or plain tap water is added to the tank water just before the evolution is conducted. A dye can be put in all of the liquids to prevent the person who adds the liquid to the tank water at the training ground from knowing which liquid is being added to the tank water. The bottle containing the liquid can be prepared in advance and numbered by someone who is not participating in the study. A study key table can be created that shows which numbered bottle contains which of the three liquids. This information is concealed from the project team, perhaps by putting it into a sealed envelope that is opened only after all of the evolutions have been completed.

At the time each evolution is conducted, the person putting the liquid into the tank water does not know which of the three liquids is being used on a particular occasion, and neither do the crews—both the investigators and the participants are blinded to which type of independent variable is being used. This type of study design is called a *randomized, double-blind, placebo-controlled design*. If it is done correctly, it offers the best protection against bias from the participants and investigators.

A randomized, double-blind, placebo-controlled design may not always be possible. The comparison of 1½-in. (38-mm) hose to 1¾-in. (44-mm) hose on a preconnect line, for example, does not lend itself to this type of experimental design because you cannot effectively blind the crews to the size of hose they are using.

Statistical Testing

Once the data have been obtained for all groups, the hypothesis behind your improvement project or research study has to be tested. How likely are the differences you see between groups to be the result of chance rather than a real difference in the dependent variable? To answer that question, it is important to choose the right type of statistical test. Many factors must be considered in choosing an appropriate statistical test, and most of them are beyond the scope of this text. However, it is suggested that your project team seek the assistance of a well-qualified statistician (perhaps from the mathematics department of a local college) or undertake further study in this area in a statistics class or perhaps as a professional development activity by team members (Motulsky 1995). In this text, we will attempt to convey the concepts behind the statistical tests.

For the wetting-agent example discussed earlier, the statistical test should compare the data from each group (commercial wetting agent, dishwashing liquid, tap water, placebo) against the others. We want to know the probability that the differences between the groups were the result of chance and not the result of a real difference. This probability is expressed as a **P value**. A P value of 0.25 means that there is a 25% probability that the differences are the result of chance. In most studies, a P value of 0.05 or less is generally considered to provide reasonable assurance that the differences between the groups are real. To put it differently, most researchers want there to be only a 5% probability or less that differences in the data are the result of chance (random variations) before they decide that the change really does make a difference. The smaller the P value, the more likely it is that differences between the groups are real. **TABLE 6-1** shows the results of statistical testing for differences among the wetting agents.

There is also another possibility for getting a small P value even though there isn't a real difference between the groups: There might be an extraneous variable that has not been controlled. For example, there might have been more wind during the test fires when using one of the wetting agents. If we did not recognize this as a factor that might influence the results and did not take effective steps to control it, there might be enough bias in the results to lead us to the wrong conclusion about the wetting agents. This potential is important to consider when designing a research or improvement project. It is also an important issue to consider when reading and interpreting research on performance improvement that is reported in technical and trade journals.

Statistical process control (SPC) charts, as discussed in Chapter 4, *Defining and Measuring the Problem*, can be used to tell the difference between variations arising from common causes versus those arising from special causes. When you change the process in an effort to improve its performance, you are trying to intentionally trigger special-cause variation. If your change does not trigger any statistical signals of special-cause variation (positive or negative), this is a strong indication that your change has no significant effect on the dependent variable the chart is monitoring. This represents a type of hypothesis test, but it does not return a P value.

Statistical Power

How much difference do you need to see between the process you currently have and a new process for it to be meaningful? This is a key question in deciding how many times you need to run a test or how much data you need for each group that will be compared.

Suppose your team is trying to speed up its process for establishing a water supply. The team is comparing different ways to make a hydrant hook up to a 5-in. (12.7-cm) supply line (**FIGURE 6-2**). The current process takes, on average, 200 seconds. To make it worth the bother of changing, how much improvement would your team need to see from an alternative process? If it only improved the time by 1 second (0.5%), that may not be worth the hassle. Your team may decide that if a change does not make at least a 40-second difference (20%), there is insufficient reason to change equipment, training, and so on. Therefore, you need to run the hydrant hookup test often enough to detect a difference of 20% or greater. This is what **statistical power calculations** are for: They tell you how often you need to run a test to detect at least a specified level of difference between groups.

FIGURE 6-3 shows two histograms. The histogram made up of orange blocks shows the results of 100 hydrant hookup tests using the current method, which we will call Method A. Each column on the histogram represents a 5-second time increment. Every time the hydrant hookup took 190 to 194 seconds, a green block goes into that column. Each block stacks on top of the ones that are already there. The histogram made up of yellow blocks shows the same data but with each time reduced by 20%. This lets us see what the histogram looks like for a process that is 20% faster (Method A minus 20%).

In **FIGURE 6-4**, there are three red blocks showing the time it took using a different method, Method B, to make the hydrant hookup. With just three blocks, do you get an impression of whether there is likely to be at least a 20% difference between Method A and Method B? Because the three red blocks are so far away from the centers of the orange and yellow block groupings, you probably get the impression that there will be a difference of 20% or more between Method A and Method B.

Let's try another method for hydrant hookups. Method C data are shown by white blocks in **FIGURE 6-5**. Because the white blocks are closer to the orange blocks, the improvement is harder to gauge. This simple example illustrates the key ideas in statistical power calculations.

TABLE 6-1 Times to Extinguish a Fire Using Various Wetting Agents (in seconds)

	Commercial Wetting Agent	Dishwashing Liquid	Tap Water
	342	325	329
	317	327	330
	349	327	370
	341	328	331
	352	329	327
	353	329	367
	341	329	377
	311	333	334
	346	334	345
	360	334	366
	351	336	329
	330	336	355
	326	337	375
	352	338	349
	317	344	366
	321	346	329
	329	347	369
	327	347	373
	315	348	360
	319	354	368
	314	359	351
	336	360	350
	314	361	361
	324	362	353
	316	366	346
	329	366	362
	361	367	346
	343	369	358
	354	372	334
	344	372	337
	337	374	333
Mean	334.55	346.97	350.97
Means and P Values for Differences between Wetting Agents		P Value	
Commercial wetting agent versus water		<0.0003	
Commercial wetting agent versus dishwashing liquid		0.008	
Dishwashing liquid versus water		0.5865	

When we are interested only in bigger differences between groups, we need less data to tell whether there is enough difference that it matters to us. However, if even a small difference matters to us, then we need much more data to see whether there is enough difference to matter to us.

In **FIGURE 6-6**, we now have 75 white blocks. Do you get the impression that there is at least a 20% difference between Method A (orange blocks) and Method C (white blocks)? This shows that the more data you have, the more certain you can be that there is or is not enough difference between the groups to matter. The calculations for determining statistical power can tell you how much data you need to detect a given level of difference between two groups. Or, working the calculations from the other direction, you can see how much difference your available data are capable of detecting.

FIGURE 6-2 A study might be conducted to determine the best method for connecting to a hydrant using a 5-in. (127-mm) supply line.

When There Isn't a Difference

Sometimes your project team will pick a factor to change, come up with a sound experimental design, run the appropriate statistical tests, and find that the change did not improve the process. This should be expected some of the time. Many argue that if you do not fail more often than you succeed, you are not trying hard enough to be innovative (Sloane 2004). The real message here is not to get discouraged if your change does not improve the process. Indeed, there may be others who also thought that the change you attempted would make the process better. If you publish your results in a technical or trade journal article, you can help others learn from your attempt. The key is to learn from the attempt. Reflect back on how you chose which factor to change. How could you have

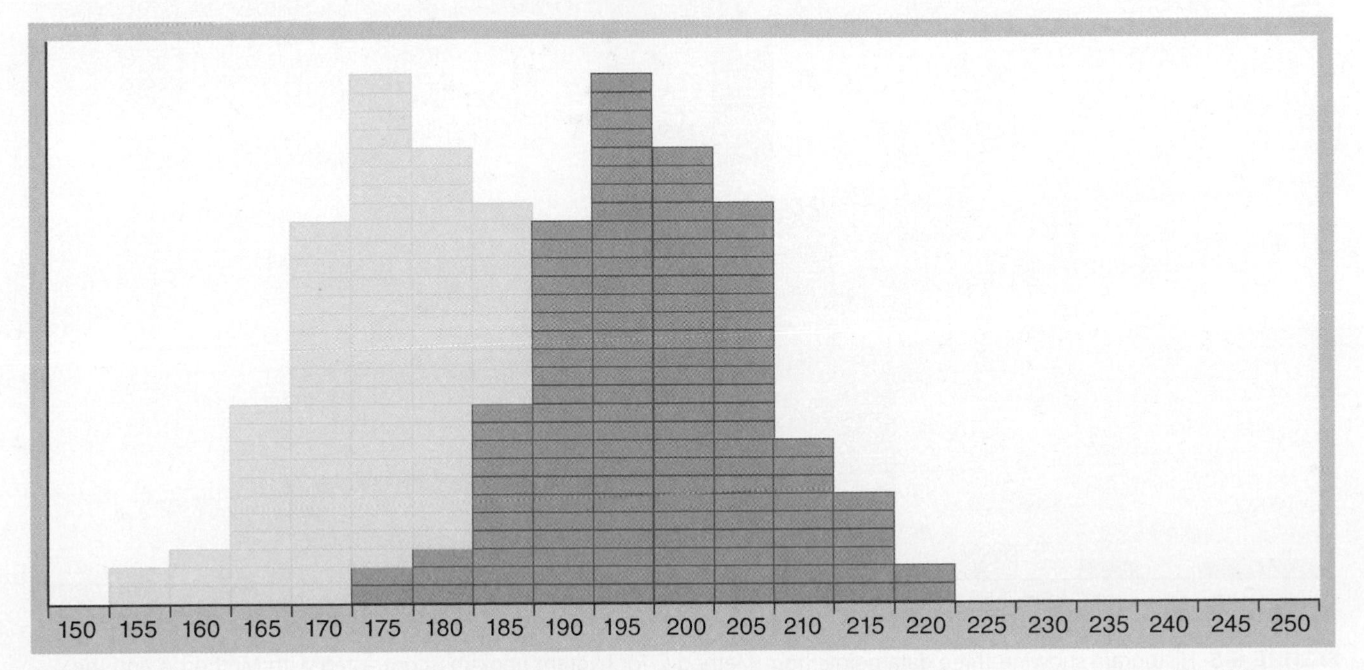

FIGURE 6-3 Two histograms of hydrant hookup times by Method A and a second histogram showing what a 20% difference looks like.

changed your selection process to choose a factor more likely to improve the process? Use your experience to try another factor to change and test. The next one may also fail to improve the process, but each time you try, your skill level will improve, you will gain deeper insights into your process, and you will be one step closer to finding what to change to make the desired improvement.

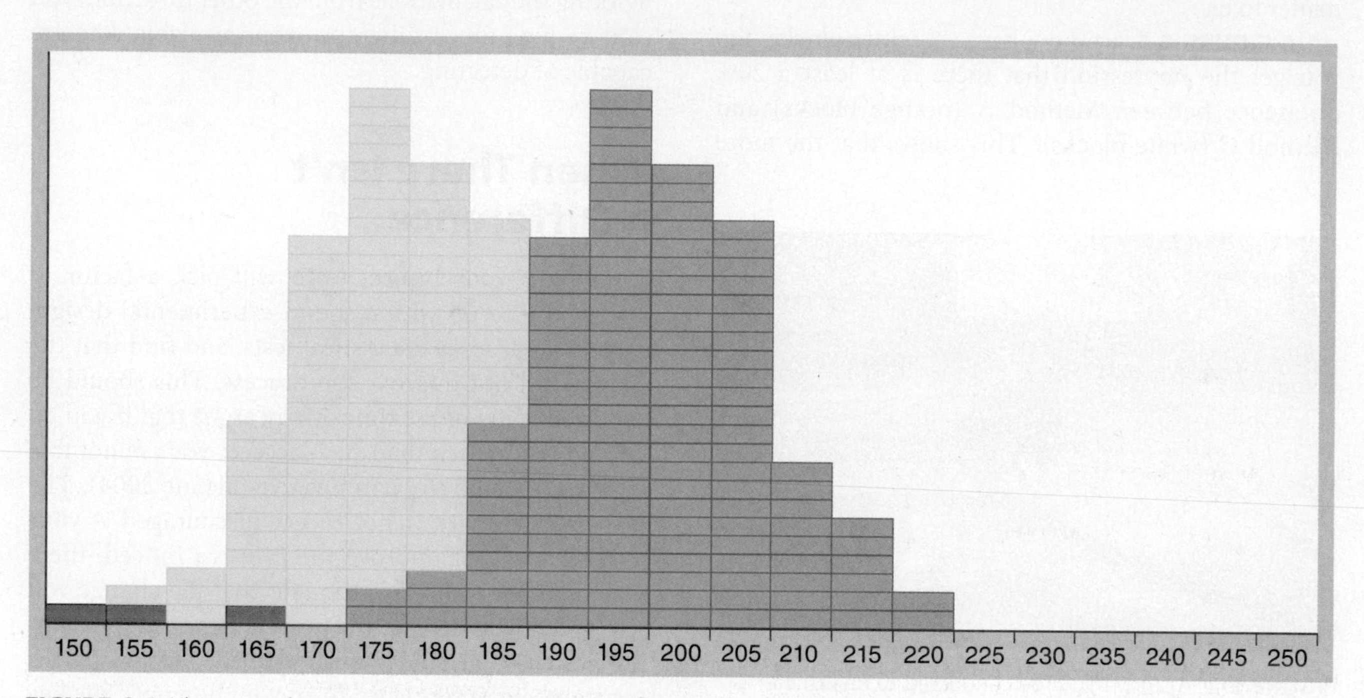

FIGURE 6-4 Histogram showing three data points from Method B for hydrant hookups, contrasted with Method A and the 20% difference example for Method A.

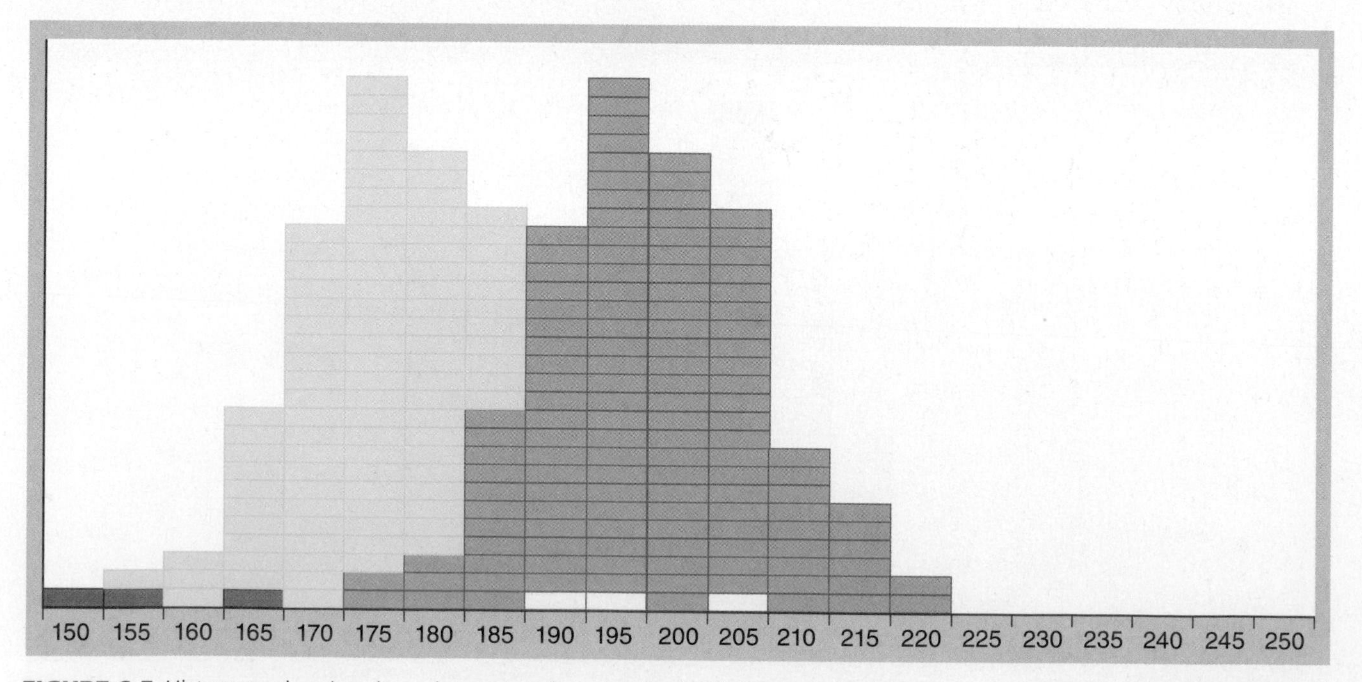

FIGURE 6-5 Histogram showing three data points from Method C for hydrant hookups, contrasted with Method A and the 20% difference example for Method A.

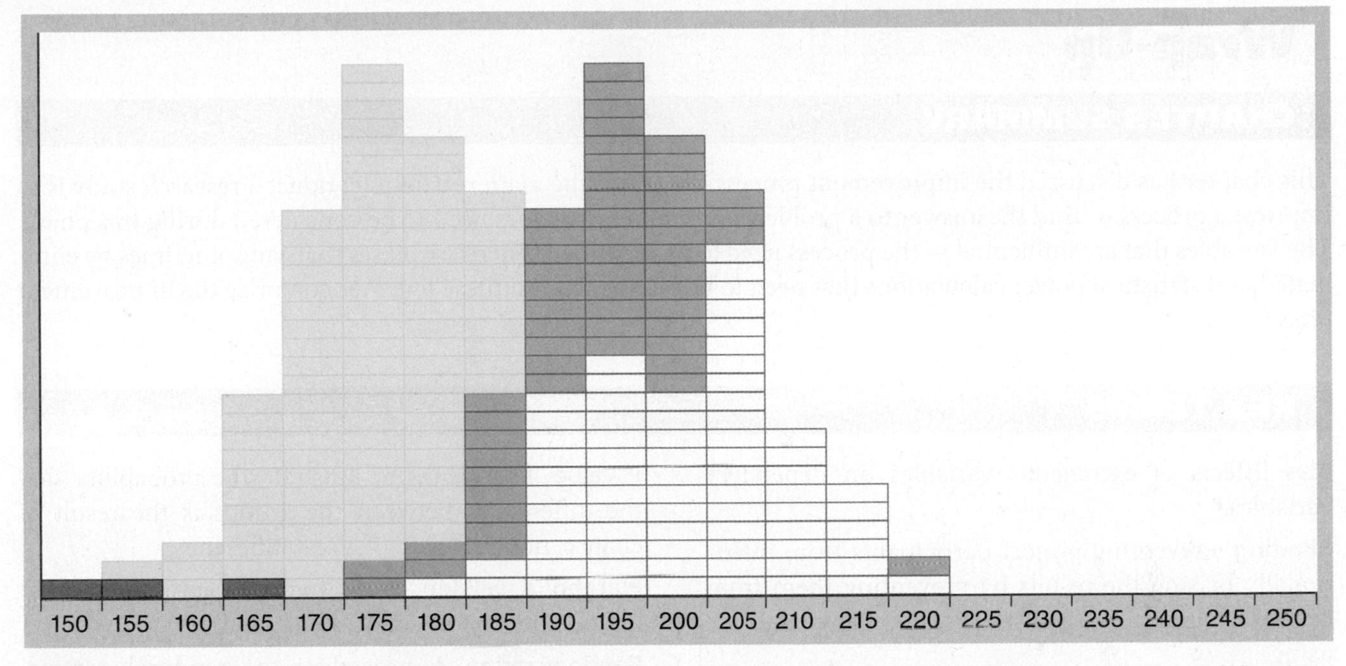

FIGURE 6-6 Histogram showing 75 data points from Method C for hydrant hookups, contrasted with Method A and the 20% difference example for Method A.

THE RESEARCH EXPERIENCE

The National Institute for Occupational Safety and Health (NIOSH) is very proactive in fire-fighter research. In early 2009, NIOSH published *NIOSH Fire Fighter Fatality Investigation and Prevention Program: Leading Recommendations for Preventing Fire Fighter Fatalities, 1998–2005*, which summarizes the most frequent recommendations from the first 8 years of the NIOSH Fire Fighter Fatality Investigation and Prevention Program (FFFIPP). It was compiled to help emergency services sector (ESS) departments and agencies protect personnel, their foremost critical infrastructure, by developing, updating, and implementing effective policies, programs, and training to prevent line-of-duty deaths (LODDs).

The report synthesizes 1286 individual recommendations from the 335 FFFIPP investigations conducted through 2005. The investigations, which involved 372 LODDs, encompassed circumstances such as cardiovascular-related deaths, motor vehicle accidents, structure fires, diving incidents, and electrocutions in career, volunteer, and combination departments in both urban and rural settings throughout the United States.

The recommendations were developed by NIOSH investigators using existing fire service standards, guidelines, standard operating procedures (SOPs), and other relevant resources. NIOSH personnel reviewed records, such as police, medical, and victims' work/training records, as well as departmental procedures, and examined the incident site and equipment used, including personal protective equipment. The Emergency Management and Response—Information Sharing and Analysis Center (EMR-ISAC) notes that for each of the 10 recommendation categories that follow, the report presents an overview of the category, category recommendations, a case example of a fatality investigation report summary, ESS department self-assessment questions, and key resources. The recommendation categories are as follows:

- Medical screening
- Fitness and wellness
- SOPs and guidelines
- Communications
- Incident command
- Motor vehicle
- Personal protective equipment
- Strategies and tactics
- Rapid intervention team
- Staffing

NIOSH Fire Fighter Fatality Investigation and Prevention Program, Leading Recommendations for Preventing Fire Fighter Fatalities, 1998–2005 (2008). Department of Health and Human Services, Centers for Disease Control and Prevention National Institute for Occupational Safety and Health.

Wrap-Up

CHAPTER SUMMARY

This chapter has discussed the improvement process. Typically, the main reason to conduct a research study is to improve a process or find the answer to a problem. A number of factors need to be considered during this phase. The variables that are influential in the process need to be identified. There are biases that can sometimes be eliminated and statistical power calculations that need to be performed. All these together comprise the improvement phase.

KEY TERMS

Bias Effects of extraneous variables on dependent variables.

Blinding Preventing project participants from intentionally biasing the results by preventing them from knowing which version of the process change they are using.

Dependent variables The measure of performance (i.e., the performance indicator) that you are trying to change.

Extraneous variables Factors, other than the independent variables, that may influence the results seen in the dependent variable.

Hypothesis The change we make to an independent variable.

Independent variables Factors that influence process performance.

Improve phrase Choosing the factor to change in an effort to make performance better.

P value The means to illustrate the probability that the differences between the groups is the result of chance, not the result of a real difference.

Placebo A version of the independent variable that has no effect on the dependent variable.

Randomization A procedure for randomly choosing which item is used each time the experiment is conducted.

Statistical power calculations Calculations tell how often you need to run the test to detect at least a specified level of difference between groups.

Statistically significant As shown by statistical tools, the differences observed between performance before the change and performance after the change (or between one group and another group) that are unlikely to be the result of chance—that is, the differences are real.

REVIEW QUESTIONS

1. Describe some of the issues or factors you would consider when making changes during the improvement phase.
2. Construct a hypothesis for an experiment involving the use of a fog nozzle versus a straight-tip nozzle at a building fire.
3. Explain what it means to randomize a study.
4. Explain the importance of a P value.
5. Describe the factors involved in bias.
6. How could you potentially prevent bias from occurring in a study?
7. Describe the improvement process.
8. Explain the difference between independent and dependent variables.
9. Explain statistical power.
10. What is the importance of a study without a significant difference?

REFERENCES

Motulsky, H. 1995. *Intuitive Biostatistics*. New York: Oxford University Press.

Sloane, P. 2004. "Failure Is the Mother of Innovation." Accessed June 15, 2020, https://innovationmanagement.se/imtool-articles/failure-is-the-mother-of-innovation/.

CHAPTER

Maintaining Control

LEARNING OBJECTIVES

Upon completion of this chapter, you should be able to:

- Explain the four components of a process control plan.
- Describe the process of transitioning responsibilities in the collection and analysis of data.
- Discuss the tools used to monitor process performance data on an ongoing basis.

Case Study

The Somewhereville Fire Department implemented a new process to capture three event times on residential structure fires:

- Interval from time on-scene to first water on fire
- Interval from time on-scene to primary search completed
- Interval from time on-scene to secondary search completed

After a couple of months of getting used to the new process, the compliance rate of crews reporting these times over the radio was over 90%. After 12 months, the rate of compliance began falling. For the third straight month, the compliance rate has dropped approximately 5% each month. At the end of this third month, the compliance rate was only 73%.

Questions:

1. If you were a member of the improvement-project team that designed and oversaw the implementation process, what issues might you have anticipated as reasons for a decline in compliance for reporting these times over the radio?
2. For one of those anticipated issues, what corrective actions would you have recommended be taken in the event that compliance rates fell?
3. Can you identify a program or process in your department that started out working well but declined in performance over time? What do you think was the reason? What might have been done to correct it?

Introduction

In the *define* phase of the project, the problems and limitations related to a specific customer need and the associated process are articulated. During the *measure* phase, the inner workings of the process are investigated and documented. The baseline level of performance of the process is also measured. In the *analyze* phase, the reasons behind the process problems or limitations are uncovered. In the *improve* phase, various ideas are tested to find workable solutions to overcome the problems or limitations in the performance of the process. In this chapter, we discuss the *control* phase, in which steps are taken to make these improvements a permanent part of the process and establish ways to monitor performance over time so that any deterioration in performance can be identified and resolved quickly.

Process Control Plan

The performance gains made in the *improve* phase are not guaranteed to continue. As the enthusiasm and energy from the performance-improvement project begin to fade, people are prone to forget exactly what they are supposed to do and exactly how to do it over the long term. A sound **process control plan** ensures "that the good improvements established by your project will not deteriorate" (iSixSigma n.d.).

A process control plan consists of four components:

- Sensor
- Alarm
- Control logic
- Validation

Sensor

The **sensor** is a permanent mechanism for measuring process performance, which consists of one or more process performance indicators that monitor the situation. Performance indicators, statistical process control (SPC) charts, and capability indexes, as discussed in Chapter 4, *Defining and Measuring the Problem*, can give you a baseline for how well a process is working. In the control phase, these same tools will be used as sensors to monitor the newly improved process performance on an ongoing basis.

Alarm

When the performance indicator results meet certain conditions, the **alarm** component of the process control system should be triggered. Two conditions should trigger an alarm: a situation that exceeds the control limit and a situation that exceeds the specification limit.

The first condition that should trigger an alarm is the presence of a statistical signal that suggests special-cause variation is taking place. This is referred to as a **control-limit alarm**. A control-limit alarm allows process operators and owners to be alerted when something unusual is taking place in the process.

The second condition that should trigger an alarm is a performance level that falls outside the upper or lower specification limit. This is called a **specification-limit alarm**.

Control Logic

The **control logic** of the process control system describes the appropriate actions to take in response to an alarm condition. If the variation that causes a control-limit alarm is that the process is performing unusually well, the root cause should be identified and studied to see whether these circumstances can be replicated to become a permanent feature and another improvement in process design.

For example, suppose that all of the engine companies are required to conduct routine fire inspections and establish preplans for low-hazard businesses in their primary response area. Crews are expected to complete inspections and preplans for all low-hazard businesses in their primary response area at least once every 3 years.

Now suppose that the members of the B-shift crew on Engine 66 change the way they conduct their inspections and are now performing at a significantly higher level than the other crews. That higher level of performance will show up as a special-cause variation in the SPC chart. After the control-limit alarm was triggered, the battalion chief and the fire marshal studied what the B-shift crew was doing differently. They discovered that the E66 B-shift crew started mapping out their schedule for inspections by finding addresses in the same area and along the same side of the street. Other crews were conducting inspections in alphabetical order. The new process allowed the E66 B shift to reduce travel time between inspection sites, thereby getting more inspections completed within the same time frame. This "best practice" was shared with other crews. Further, the procedures for conducting the inspections were changed to match E66-B's process, with the result that now it works more efficiently across the entire department (**FIGURE 7-1**). Once all the crews are accustomed to using the new process for mapping out their low-hazard inspections, the control limits on the SPC charts can be recalculated, and monitoring at the new level of performance will carry forward.

If the variation that triggered the control-limit alarm was unfavorable, the situation should also be studied to identify the root cause. Once the root cause is identified, steps can be taken to prevent it from occurring again—or at least to mitigate the negative effects.

FIGURE 7-1 When one crew has a higher productivity rate than other crews, the process the crew is using needs to be observed to determine how to improve the process.

As mentioned earlier, when the performance level falls outside either the upper or lower specification limit, it should trigger a *specification-limit alarm.* For a one-tailed process, such as an emergency response time that needs to be as short as possible, there will be only one specification limit.

If the specification-limit alarm does not occur in conjunction with a control-limit alarm, the process is operating normally. This just tells you that the normally operating process is not capable of operating consistently within the specification limits. The design of the process needs to be changed to make it capable of operating within the specification limits.

In contrast, if the specification-limit alarm does trigger together with a control-limit alarm, this indicates that something occurred that was sufficiently different to represent unusual process behavior and that there was not enough latitude between the control limits and the specification limits to absorb the special-cause variance and still stay within the specification limits. If the special-cause event is very rare, no action may be warranted. If it is more common and cannot be effectively prevented or mitigated, a redesign of the process to provide more room to spare within the specification limits may be needed to solve the problem.

The various combinations of alarm conditions and appropriate responses are summarized in **TABLE 7-1**.

TABLE 7-1 Appropriate Responses to Specification-Limit and Control-Limit Alarms

Alarm Condition	Process Status	Control Logic/Remedy
Control-limit alarm only	Process shows unusual behavior but is still operating within specifications.	Find root cause; if positive, try to incorporate it into process design; if negative, try to prevent or mitigate recurrence.
Control-limit alarm and specification-limit alarm	Because process performance is falling outside the specification limits, the control-limit alarm is most likely negative.	Find root cause; try to prevent or mitigate recurrence. Consider process design change so that future special variations can be tolerated without violating specification limits.
Specification-limit alarm only	Because the process is behaving normally (no special-cause variation), the current process design is not capable of performing consistently within specifications.	Redesign the process to perform consistently within specifications.

Validation

Once the appropriate actions have been taken from the standpoint of control logic, the new level of process performance should be documented. This is the **validation** component of the process-control system. Validation may be accomplished with a continuation of the SPC chart to show the new level. It could also be documented with a capability index, such as Cpk (see Chapter 4, *Defining and Measuring the Problem*).

FIGURE 7-2 During an improvement project, members of the project team need the time and resources to collect data and analyze the data using statistical process control (SPC) charts.

Transitioning Responsibilities

During an improvement project, members of the project teams are given the time and resources to collect data and analyze the data using SPC charts (**FIGURE 7-2**). In the control phase, the responsibility for continuing to collect data and performing SPC analysis should shift from the project team to the people who will have ongoing responsibility for the process—the process "owners." To make this a permanent, routine part of how the process operates, a way to collect data and put the data into the SPC charts should be embedded in the routine activities of the process owners.

Many fire departments use software tools for their fire incident reporting and records management systems. Ideally, the data needed for ongoing monitoring of the process performance indicators are collected by these tools. Calculation and display of the performance indicators are also ideally done by these software tools so that the whole process of data collection, analysis, and reporting is automated (**FIGURE 7-3**). However, most of these tools have not yet evolved to

FIGURE 7-3 Using an automated process for monitoring and reporting how well your processes are performing across your department will increase your ability to monitor and improve your system.

this level of sophistication. If an automated process for monitoring and reporting how well processes are performing across your entire department is something that your department would like to have, start asking your software supplier for it. In the meantime, manual processes will have to be put into place.

Study how the project team collected, analyzed, and reported the performance indicator(s) for the dependent variable(s). Then consider how their methods can be modified to do this over the long term. Answering the following questions will help to determine these variables:

- Which staff position inside the process is best able to do the data collection?
- Which staff position is best suited to move the data into the SPC software?
- How will process operators and owners be trained to interpret and take appropriate actions based on the SPC charts?
- How will the updated SPC charts be shared within the group of process operators and the managers who function as process owners?
- How will other managers be able to access the SPC charts?

For more frequent processes, such as emergency responses, the calculations and reporting may be needed daily, if not more often. You will want to know sooner rather than later if the process for emergency responses is starting to behave in an unusual fashion. For a slower-moving performance indicator, such as employee turnover, data collection and calculation might be done monthly.

How the performance indicator results are reported is another important consideration. As your organization builds up its library of performance indicators over time, it will be very helpful to display them in a format that allows process owners and upper management to see, at a glance, how all of the monitored processes are working. This allows managers to have an "early warning system" that will let them know if things are getting off-track. This is the idea behind a **management dashboard system** (Wikipedia 2020). The results from manual or automated performance indicator calculations can be entered into a good management dashboard system.

Wrap-Up

CHAPTER SUMMARY

The define phase of a project includes identifying the problems and limitations related to a specific customer need and the associated processes. The measure phase involves investigation and documentation of the inner workings of the process, as well as the measurement of the baseline level of performance. In the analyze phase, the reasons behind the process problems or limitations are uncovered. In the improve phase, various ideas are tested to find workable solutions to overcome the problems or limitations in the performance of the process. In the control phase, steps are taken to make these improvements a permanent part of the new process and establish ways to monitor performance over time so that any deterioration in performance can be identified and resolved quickly, such as with the use of a process control plan. The combination of these efforts allows a department to methodically take an issue and determine how best to correct or improve the situation.

KEY TERMS

Alarm When the performance indicator results meet certain conditions, the alarm component of the process control system is triggered.

Control-limit alarm The presence of a statistical signal that suggests special-cause variation is occurring.

Control logic The component of the process control system that describes the appropriate actions to take in response to an alarm condition.

Management dashboard system A library of performance indicators displayed in a format that allows process owners and upper management to see, at a glance, how all of the monitored processes are working, thus allowing managers to have an "early warning system" that will let them know when things are getting off-track.

Process control plan A way to ensure that favorable improvements made by a process-improvement project will not deteriorate over time.

Sensor The component of the process control system that serves as a permanent mechanism for measuring process performance, typically consisting of one or more process performance indicators that can monitor process performance over time.

Specification-limit alarm An alarm that is triggered when the performance level falls outside the upper or lower performance specification limit.

Validation The component of the process control system that documents the new level of process performance after appropriate actions have been taken from the standpoint of process control logic.

REVIEW QUESTIONS

1. List the four components of a process control plan.
2. Explain the idea behind a management dashboard system.
3. Explain the difference between the responsibilities of those who collect the initial data and those who maintain the data.
4. Describe the process of transitioning responsibilities for the collection and analysis of data.
5. Discuss the tools used to monitor process performance data on an ongoing basis.
6. Describe validation.
7. What is a specification-limit alarm?
8. 8–10. Complete the following table:

Alarm Condition	Process Status	Control Logic/Remedy
Control-limit alarm only	Process shows unusual behavior but is still operating within specifications.	8.
9.	Because the process performance is falling outside the specification limits, the control-limit alarm is most likely negative.	Find the root cause; try to prevent or mitigate recurrence. Consider process design change so that future special variations can be tolerated without violating specification limits.
Specification-limit alarm only	10.	Redesign the process to perform consistently within the specifications.

REFERENCES

iSixSigma. n.d. "Process Control Plan." Accessed June 15, 2020, https://www.isixsigma.com/dictionary/process-control-plan.

Wikipedia. 2020. "Dashboards (Business)." Accessed June 15, 2020, http://en.wikipedia.org/wiki/Executive_dashboard.

CHAPTER

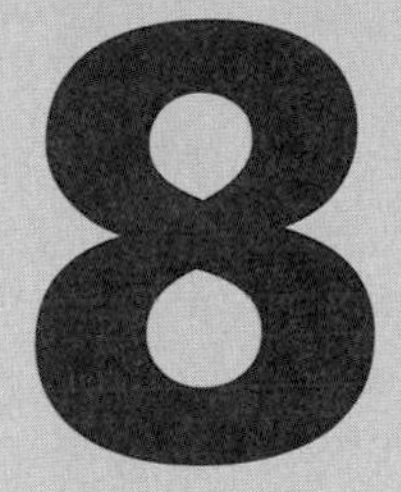

Qualitative Research

LEARNING OBJECTIVES

Upon completion of this chapter, you should be able to:

- Describe qualitative research.
- Discuss the various types of studies used in qualitative research.

Case Study

The staff of the fire prevention division, working in collaboration with its fire prevention improvement team (as described in the case study at the end of Chapter 6, *Improving the Process*), identified through a Pareto analysis that the leading site of origination for residential structure fires was the kitchen. In 60 percent of those fires, there were no smoke alarms.

1. What type of qualitative research study would you suggest for determining why those homes did not have smoke alarms? Why?
2. The fire prevention improvement team was interested in knowing the level of importance that citizens in their community placed on having smoke detectors in their homes. What type of qualitative research would you suggest to gain insights on this issue?

Introduction

Quantitative research methods are used to gather data and test a hypothesis for how changes in a process may resolve a problem or improve fire department performance. Quantitative research is very effective in processing large volumes of data and then making generalizations about the topic under investigation using statistics. In contrast, qualitative research is best suited for the job of telling a story from the point of view of the subject or as an observer. The results of qualitative research help us to get the "big picture" and put the issues, problems, challenges, and opportunities for improvement into an appropriate perspective (**FIGURE 8-1**).

FIGURE 8-1 Qualitative research is used to get the "big picture" about a problem at hand. It can involve many different questions you want to answer.

Qualitative Research

Qualitative research was developed primarily for use in the social sciences. In that arena, it sought to answer questions about human behavior, culture, attitudes, and the effects of events on people. It has evolved into a robust set of methods that seeks to explain things as they actually occur. Unlike the controlled experimental method of quantitative research, qualitative research usually does not try to change independent variables or limit the effect of confounding variables. The data coming from qualitative research are primarily subjective.

In terms of its utility for fire research, qualitative research can be the right approach to find answers to questions about why things are the way they are or why people and communities do what they do. Qualitative fire research may ask questions such as the following:

- Why do some of your fire fighters have low morale?
- What impressions do various groups in your community have about the quality of services provided by your fire department?
- Why do citizens in your community wait so long before calling for help when they are having a heart attack?
- Why are the rates of utilization of smoke detectors and carbon monoxide detectors so low in your community?
- How do people in your community decide when and where to evacuate in the face of an oncoming hurricane? Are the reasons the same for different groups, such as the elderly, those with special medical needs, the affluent, the poor, or immigrant populations?

FIGURE 8-2 It is valuable to improve our understanding of the whole rather than of each component.

One of the general ideas behind qualitative research is that there is value to be gained by improving our understanding of the whole, rather than studying each of the parts separately. For example, understanding the complexities of interactions between people and processes within a fire department, a community, or a disaster scene is extremely valuable when you are trying to be a better manager and leader who has responsibilities for developing policies and making strategic and tactical decisions (**FIGURE 8-2**).

There are several broad categories of qualitative research. Each has different applications and strengths. It is the responsibility of the researcher to determine what the research goals are and then select the type of research that is best suited for meeting those goals.

Ethnographic Studies

Ethnography is a qualitative research method used for descriptive studies of culture in specific populations or groups that share a common parameter, such as their profession, nationality, religion, location, or even a having been exposed to the same event or experience. An ethnographic fire research study might explore how nationality and immigration status influence the response of Vietnamese immigrant families to your department's current methods for community education in fire and injury prevention.

An ethnographic study design is built around plans for conducting interviews and making observations of behavior. The starting point for planning the study is clearly stating the research question. From this, the researcher can prepare a list of questions for use in a formal or informal interview process. The responses to questions are most often recorded in notes taken by the researcher during the interview. This may be supplemented by audio or video recording of the interview. The advantage of audio or video recording is that it provides an opportunity to replay the response to make more accurate and detailed notes. It may also allow the researcher to focus on the subject (the person being interviewed) rather than having the distraction of note taking. Video recording offers the opportunity to replay responses to take note of nonverbal responses, such as changes in body language.

A formal interview process is structured. It attempts to ask each subject the same questions and record their responses to each item. This reduces the potential for variation in responses due to the way a question was asked.

An informal interview process is unstructured. General themes and "starter" questions are used to help the researcher touch on all of the key issues during the interview. The interview is more conversational and therefore gives the researcher more latitude in taking the interview in different directions depending on the responses of the subject.

An ethnographic study may also involve observation of subject behavior. Most commonly, it is akin to being "a fly on the wall," where researchers try to position themselves in a way that minimizes the impact of their presence on the behavior of the subjects. This is much easier to do with larger groups in public spaces than with individuals or families in their homes.

Phenomenological Studies

Phenomenological research is used to study a phenomenon. The phenomenon studied in a fire research project might be a fire, disaster, training evolution, public education class, or an assessment center for selecting new fire officers (**FIGURE 8-3**). The purpose of

FIGURE 8-3 An assessment center might be one type of setting for a phenomenological study.

a phenomenological study is to improve your understanding about something that exists in the world we live in. It should fill a gap in your knowledge. It may not provide conclusive answers about why something happens or why people react the way they do, but it can help clarify and illuminate your understanding of the phenomenon.

Phenomenological studies are primarily observational. Planning for this type of study should include considering ways that observations can be made without influencing subjects (as with the idea of the fly on the wall). Planning should also include the development of data collection tools that address key points or issues that need to be covered during the observations. It may also include a way to count how often during the study specific things are observed. Also, as described in the ethnographic methods section, audio or video recording may be used. Depending on the phenomenon, other types of data may also be collected.

Grounded-Theory Studies

A **grounded-theory study** is an extension of a phenomenological study. It complements the subjective observations with the proposal of new theories to explain what has been observed. These theories develop as the researcher simultaneously collects and interprets data during the study. As ideas for theories begin to develop, the researcher may alter the observation patterns or interview questions to see whether the idea or fledgling theory applies to other subjects. This loop of data collection and analysis is referred to as *constant comparative analysis.* The ideas are compared with the incoming data and are constantly being revised to better explain the findings. The new theories are based on or *grounded* in the data from the observations and interviews.

A good example and description of the grounded-theory method comes from the study of people's reactions to death and dying. Its development is the now widely accepted theory of how the grieving process operates (Kübler-Ross 1969).

A qualitative study on people's behavior in reaction to the death of a loved one led the researchers to the idea that there was a pattern in subject reactions. They shifted the focus of their observations to looking for this same pattern in new subjects. Adjustments were made in their ideas for the pattern and what to look for using the constant comparative analysis technique. At the end, there was a new theory and knowledge now commonly referred to as the "grieving process," with associated names for the various stages of grieving—denial, anger, acceptance, and resolution. This facilitated other new knowledge that has improved our understanding of "abnormal" grieving, which led to the development of interventions to help.

Case Studies

Case studies are used to describe a single unit—an event, occurrence, situation, organization, or individual. A case study can be used to capture as much detail and depth as possible about the subject in ways that other methods may not afford. The complexity of a case study may range from a single event of short duration to the study of a convoluted situation in a large geographically dispersed organization (such as a large urban fire department) that takes place over an extended period of time. The level of complexity will strongly affect how the study is planned for and executed.

A case study method may be applied to fire-related topics, such as an innovative new practice for recruiting volunteer fire fighters, following a single individual over the course of her career in and out of the fire service, a line-of-duty death or injury incident, a disaster, or a richly detailed examination of a single-story residential structure fire (**FIGURE 8-4**). The detail of well-done case studies can lead to more specific ethnographic or phenomenological studies, inspire new ideas for grounded-theory study, or suggest a hypothesis for a quantitative study.

A potential weakness of the case study method is that the observations from one unit may not be

FIGURE 8-4 A structure fire might be one type of case study you want to explore.

reflective of or generalize to other similar units. The way something happened in one instance may not tell you anything about what happens in other, similar instances. However, the details revealed in the one instance may provide ideas for what to look for in other instances. It is the responsibility of those who read the case study to determine whether the instance has sufficient commonality with their own case or circumstance.

The case study method is unrivaled for the study of something that is recognized as unique. The observations and lessons from the one instance may provide invaluable aid should a similar instance ever arise.

Planning for conducting a case study revolves around devising methods for effective and efficient data collection in as many ways as appropriate. This may include the use of multiple observers, electronic data-collection instruments (including audio and/or video), and multiple interviews of those involved in the subject of the study.

Data Analysis

Most of the data collected in a qualitative study will be in formats that are unstructured and laborious to process. Imagine having hour after hour of video, page upon page of notes, CDs full of photographs, and volumes of interview transcripts. The goals of the study will guide the specific methods used to digest the data, but there are some general principles common to almost all qualitative data analyses.

One of the major goals in the analysis of qualitative data is determining whether there are patterns or recurring themes within a unit that are found in common among multiple subjects. In reviewing transcripts or recordings of an interview, making notations using a set of key words or phrases that evolve from the data is useful. This is called a *coding frame.* In some projects, you may set up the coding frame ahead of time, based on what you want to look for. In other types of projects, you may need to create a list of key words and phrases as you go. In either case, as the analysis moves through the data, the list of those key words or phrases can be expanded. Sometimes two similar key word phrases may be consolidated, so the list gets shorter. The data and notes may be given multiple passes, each time refining, expanding, and consolidating.

Transcribing audio or video recordings into text can be helpful, but it is usually very time-consuming. It can also be expensive if the work has to be outsourced. Some software programs may be able to transcribe audio into text, but their accuracy should be tested on a sample of the audio. The quality of the audio recording is a major factor in the ability of the software to make a passable transcription.

As themes or patterns emerge, other data may be used to provide examples of the patterns in words, audio, video, and photographs, along with confirmations and correlations to data from other subjects or units. These patterns and themes may become the basis for conclusions or the primary results of the study.

Bias

Given the subjective nature of qualitative analyses, there is an increased potential for bias in data analysis and interpretation. If the investigators personally favor or expect a certain outcome for the study, that can potentially influence how they determine what information is important to include, or disregard, in their selection of information gleaned from the raw data.

The intentional use of bias in research is an ethical problem that investigators should be aware of and take steps to safeguard against in their work. Unintentional bias is a more difficult problem because it can manifest without the investigators even being aware that it is occurring. In either case, steps to protect against bias may be accomplished by simply being aware of its potential and having one's work reviewed during the course of a study by collaborators or even by colleagues who are not a part of the project. Readers should be vigilant for evidence of bias in their review of research.

Hybrid Studies

In the real world, there may not be clear lines of separation between the different types of qualitative studies. A real project may be a **hybrid study** that uses some parts of ethnography, other parts from phenomenology, some elements of grounded-theory and constant comparative analysis, and some parts of the case-study method. This may also include the collection of quantitative data and statistical analysis to characterize the data. A hypothesis can emerge, which gets tested. All of this is perfectly legitimate if it suits the objectives of the project. The point here is not to let these labels and categories impose unnecessary restrictions on a project. Be aware of all the different qualitative and quantitative methods. Mix and match them to best answer the questions you are asking.

Wrap-Up

CHAPTER SUMMARY

Qualitative research is used to answer questions that are subjective in nature. Depending on the purpose of your study, you will decide the type of study you should use for your research topic. Qualitative research can provide in-depth answers to questions that you may earnestly seek to answer.

KEY TERMS

Case studies Studies that describe a single unit, which may be an event, occurrence, situation, organization, or individual.

Ethnography A qualitative research method used for descriptive studies of culture in specific populations or groups that share a common parameter, such as their profession, nationality, religion, location, or even having been exposed to the same event or experience.

Grounded-theory study An extension of a phenomenological study.

Hybrid studies A study that combines ethnography, phenomenology, grounded-theory and constant comparative analysis, and the case study method, and may also include the collection of quantitative data and statistical analysis to characterize data.

Phenomenological research Research that aims to improve our understanding about something that exists in the world we live in.

Qualitative research Developed primarily for use in the social sciences as a method to answer questions about human behavior, culture, attitudes, and the effects of events on people.

REVIEW QUESTIONS

1. Explain qualitative research.
2. Write three examples of questions that you could use qualitative research to answer. Do not duplicate any already mentioned in this chapter.
3. Compare and contrast ethnography studies, grounded-theory studies, case studies, and phenomenological studies. Include when it is appropriate to use each type of study.
4. Describe how to utilize the data gathered from a qualitative study.
5. Explain the purpose of a hybrid study.
6. What type of study would be best used for a research project involving the number of Hispanic females in the fire service?
7. What type of study would be best used for a research project to analyze a major building fire in your jurisdiction?
8. You have just implemented a new procedure on initial fire attacks. What type of study would be best used for a research project to determine how personnel feel about the change in policy?
9. What type of study would be best used for a research project involving children hearing and responding to a smoke detector in the home?
10. Explain the data analysis process of a qualitative study.

REFERENCES

Kübler-Ross, E. 1969. *On Death and Dying.* New York: Simon & Schuster.

CHAPTER

Ethical Considerations

LEARNING OBJECTIVES

Upon completion of this chapter, you should be able to:

- Explain the importance of conducting research in an ethical manner.
- Discuss the purpose and process for an institutional review board (IRB) review and approval.
- Discuss the issues surrounding a research project that might cause an ethical dilemma.

Case Study

The Somewhereville Fire Department heard about a new way to prevent building-to-building fire spread by applying a chemical solution to building exposures. The chemical is not currently being sold commercially for this purpose.

1. What information would you want to know about the chemical before considering its use in your department's fire-ground operations?
2. You later discover that the chemical is commonly used in food and is generally considered safe. Are there any other issues or ethics to consider before trying it in fire-ground operations?

Introduction

The scope of research and process-improvement projects that a fire service professional may be engaged in is vast. Inevitably, these projects will involve individuals and organizations that have interests that may not be the same as those of the project or project leader. Some of these differences need to be addressed from an ethical perspective.

Research in the Fire Service

Research and process-improvement projects are needed to make advances in how fires are prevented and extinguished, how fire-fighter and victim safety is protected, and how fire-ground injuries are prevented and treated (**FIGURE 9-1**). Although these projects may provide many benefits to citizens, fire fighters, fire departments, communities, and society overall, they must be balanced against preserving the interests of the persons and organizations that may be put at risk by their participation. Those risks may be physical, psychological, legal, economic, or social.

How are the interests of fire fighters protected in a research project about trying to find better types of safety equipment? How are the interests of a fire department protected in a study that uses its data in comparison with data from other departments? How are the interests of property owners protected in a study seeking to develop better fire combat tactics? How are the interests of fire victims protected in a study evaluating a new method for detecting exposure to poisonous gas?

Medical research has developed a rich body of knowledge and processes for addressing ethical issues in research and process improvement. It provides a very useful and relevant example for how to address ethical considerations in fire research and process improvement.

FIGURE 9-1 Research is needed in the fire service to make advances in how fires are prevented and extinguished, how fire-fighter and victim safety is protected, and how fire-ground injuries are prevented and treated.

The Belmont Report

A landmark project was conducted in 1974 by the National Commission for the Protection of Human Subjects (National Institutes of Health n.d.). The result was called the **Belmont Report** (National Commission for Protection of Human Subjects of Biomedical and Behavioral Research 1979). It provided a major update to the ethical standards for studies involving human subjects. It continues to serve as the basis for all federal regulations pertaining to research involving human subjects. Three guiding principles were articulated in the Belmont Report: respect for persons, beneficence, and justice.

Respect for Persons

The ethical principle of respect for persons in research recognizes that people have the right of choice to participate, or not, in a research project. This right of choice assumes that they have the mental and legal capacity to make an informed choice. *Informed choice* means that potential participants are fully informed of any risks associated with their participation.

People who do not have the mental and legal capacity to make an informed choice need extra protection of their interests. Lack of mental capacity pertains to those who have cognitive limitations such as mental impairments, dementia, some types of mental illness, and other conditions in which the comprehension needed to give informed consent is difficult or questionable.

Lack of legal capacity may refer to the principle of respect for persons, including children and fetuses. It also may include those in subordinate positions who may be subject to coercion, such as students, employees, and prisoners.

The vulnerability of a fetus and its inability to express any preference or concern on its own behalf justifies the need for special consideration and protection. Children do not have the legal standing on their own to enter into agreements to participate in research studies because they may not fully comprehend the implications of their participation.

Children, along with employees, prisoners, and others, may be in subordinate positions in which they may be subjected to pressures from those in authority over them to participate (**FIGURE 9-2**). This pressure, if intentional, may constitute coercion. The pressure may be unintentional but may still interfere with free choice. Imagine a fire fighter being asked by his fire chief or another senior officer to participate in a research study or process-improvement project that carries some element of personal risk. The fire fighter may have concerns about how his response might influence the opinion of senior officers toward him. Prisoners may fear repercussions from prison staff for failing to consent to participate. These types of issues need to be addressed.

FIGURE 9-2 Children, among other members of society, are protected from participation in research studies. Any study should take the participation group into special consideration.

Courtesy of Estero Fire Rescue.

Beneficence

The magnitude of potential benefit resulting from a research or process improvement project should offer some level of justification for the physical, psychological, social, legal, or economic risks it poses to participants. This is the ethical principle of **beneficence**. There is a risk–benefit ratio that needs to be considered. A study that has only minor potential benefits but has significant risk would not be judged favorably from a beneficence perspective.

Justice

Some projects may be considered in which study participants receive little or none of the benefits that may come from the project. This may manifest in a study in which one group bears the risks associated with participation while another group gets the benefits.

The classic example of a study design that was lacking from the ethical perspective of justice was the syphilis study conducted in Tuskegee, Alabama, in the mid-20th century by the U.S. Public Health Service. This study examined the natural progression of untreated syphilis. The researchers specifically recruited rural black males because they could be more easily recruited with inexpensive enticements. In this case, it was the offer of free medical examinations. The results of the study helped to improve the diagnosis and treatment of syphilis. The problem is that the disadvantaged status of the study population was exploited. This group bore all of the risks for the benefit of those from any socioeconomic or ethnic category. The ethical principle of justice asserts that if all groups may benefit, then all groups should bear a fair share of the risks.

Processes for Review and Approval of Research Projects

The ethical principles outlined in the Belmont Report are the foundation for various government regulations that guide the conduct of research involving human subjects. At the federal level, these are found in Title 21, Chapter I, Parts 50 and 56 of the U.S. Food and Drug Administration (FDA) regulations. They are also found in Title 45, Part 46, of regulations from the Office of Human Research Protection. These regulations apply specifically to research projects funded with federal funds, but they probably still apply when the study is paid for with funds from other sources, such as corporations or private foundations. In general, these federal regulations are widely recognized and adopted, by law or voluntarily, by most funding agencies. They are widely recognized as ethical standards by the general scientific and academic communities.

The full range of issues that need to be addressed to be in compliance with these regulations is beyond the scope of this chapter, but the key component from a fire research standpoint is the review, feedback, and approval of proposed projects by an **institutional review board** (IRB). These bodies are also referred to as *ethical review boards* (ERBs) and *independent ethics committees* (IECs). In this chapter, we will use the term *institutional review board (IRB)*. Regardless of the name used, it is a committee that reviews applications for research projects with the goal of protecting the rights and interests of human subjects.

The IRB will review the protocols for the research project, along with the materials used in the recruitment and consent processes. If study subjects are going to be compensated in any way, those terms are also reviewed. These materials, along with the responses of researchers to questions from the IRB, will be used to determine whether the study adequately addresses the principles of respect for persons, beneficence, and justice.

The IRB may approve a study as presented, or it may flatly reject it. More commonly, the IRB will suggest changes that can be made in the study design that will facilitate its approval. In the end, the review and feedback from the IRB to the investigators usually makes for a much better study design from a scientific and methodology standpoint as well as from an ethical perspective.

There are two significant gray areas regarding IRBs pertaining to fire research and process-improvement projects. The first is that although many types of fire research involve human subjects, the studies may not be in the realm of medical or social science research—which are the disciplines these regulations are directed toward. Yet, a fire research or process improvement may pose risks to fire fighters, occupants, property owners, and their insurance companies. There may not be a legal requirement for IRB review and approval in some cases, but this is something that warrants legal review. However, from an ethical and professional perspective, regardless of a legal requirement, the risks and principles seem to be appropriate to apply to fire research studies.

Second, there is an often vague line between a research project and a process-improvement project. A quantitative research project and a process-improvement project may both use the scientific method to test a hypothesis. Both may involve human subjects and pose elements of risk to human subjects. Pure "research projects" are often conducted as an academic activity seeking to advance the science and body of knowledge in a professional discipline. The end result of a pure research project is to have the results accepted by a scientific/professional venue as a publication, scientific presentation, or lecture. As a result, many of these venues require that any research projects published and/or presented must be approved by an IRB or granted a waiver for approval because the IRB (not the researcher) determined that the project did not meet criteria, requiring their formal review. In contrast, a process-improvement project is usually conducted without the intent of external presentation in a publication or at a scientific session or professional society lecture; therefore, those requirements are not imposed.

Presently, the fire service has not evolved to the point where formal research and process-improvement studies are commonly conducted (**FIGURE 9-3**). As a result, these issues remain poorly defined and unresolved. Part of the reason for raising these issues in this chapter is to get these issues discussed in fire science classrooms at colleges and universities and in institutions such as the U.S. Fire Administration, the National Fire Academy, and the National Fire Protection Association. These issues should be discussed by groups such as the International Association of Fire

FIGURE 9-3 Presently, the fire service has not developed to the point where formal research and process-improvement studies are commonly conducted.

Chiefs, the International Association of Firefighters, the National Fallen Firefighter Foundation, the faculty and students in the Executive Fire Officer Program, and those convened in future sessions for updating the National Fire Service Research Agenda. In the meantime, it will be the responsibility of fire service professionals to become educated regarding these issues and to apply the principles as applicable and appropriate. Hopefully, those experiences will be shared in trade journal articles and at local, national, and international fire service conferences so that the art and science of fire research and process improvement can evolve to a higher level of professionalism and scientific sophistication.

Consent

If a study does involve human subjects and there are informed consent issues to be addressed, IRBs commonly require that participants sign some type of consent form. Whenever possible, a consent form should be executed before participants are included in the study. The form should inform the participant about the following:

- The goals, potential benefits, and potential risks of participation in the study
- What will happen if the subject chooses not to participate
- How the information obtained will be used
- What compensation they will receive, if any, for their participation

Sometimes, it may not be practical to get consent from potential participants beforehand. In those cases, consent may be obtained afterward if the study was observational or if there were no substantial risks or consequences to their participation regardless of which study group they were part of.

Ordinarily, if the study did pose a potential risk to participants and prior consent could not be obtained, the study would be not be conducted. However, in 1996, the FDA and the U.S. Department of Health and Human Services (DHHS) put rules into place to permit an exception to the requirements for prior informed consent for patients in research on emergency treatments (DHHS 1996). The prior consent by individuals is replaced with a study of the "acceptability" of the study to the community in which the study will be conducted. A cross-section of the community is educated about the study and its willingness to condone it. This type of consent remains controversial (Baren, Anicetti, Ledesma, Biros, Mahabee-Gittens, and Lewis 1999; Flynn 2008).

Wrap-Up

CHAPTER SUMMARY

Research in the fire service is still not as prevalent as in some other fields. Regardless, it is critical that any human subjects included in a research study should be treated in a manner that does not harm them or subject them to conditions that are not humane or safe. Ethical issues must be considered, and the proposed study needs to go through the proper vetting channels to ensure that consideration has been given to the subjects of the study.

KEY TERMS

Belmont Report The basis for all federal regulations pertaining to research involving human subjects.

Beneficence The magnitude of potential benefit resulting from a research project or process improvement project should offer some level of justification for the physical, psychological, social, legal, or economic risks it poses to participants.

Institutional review board A committee that reviews applications for research projects, with the goal of protecting the rights and interests of the human subjects who are involved.

REVIEW QUESTIONS

1. Explain the purpose of the Belmont Report.
2. Explain the purpose of an IRB.
3. Describe the process of sending a proposal through an IRB.

4. Explain the importance of conducting research in an ethical manner.
5. Discuss the issues surrounding a research project that might cause an ethical dilemma.
6. Identify four groups with whom it would be appropriate to get feedback and discuss a research project involving a fire department function.
7. Explain beneficence.
8. What are the risks to consider when conducting a research project?
9. What are the two significant gray areas regarding IRBs pertaining to fire research and process-improvement projects?
10. What does the ethical principle of justice mean?

REFERENCES

Baren, J., J. P. Anicetti, S. Ledesma, M. H. Biros, M. Mahabee-Gittens, and R. J. Lewis. 1999. "An Approach to Community Consultation Prior to Initiating an Emergency Research Study Incorporating a Waiver of Informed Consent." *Academic Emergency Medicine* 6: 1210–1215.

Flynn, G. 2008. "Community Consultation for Emergency Exception to Informed Consent: How Much Is Enough?" *Annals of Emergency Medicine* 51: 416–419.

National Commission for Protection of Human Subjects of Biomedical and Behavioral Research. 1979. *The Belmont Report: Ethical Principles and Guidelines for the Protection of Human Subjects of Research* (OPRR Publication 887-806). Washington, DC: Office for Protection from Research Risks.

National Institutes of Health. n.d. "Office of Human Subjects Research Protections." Accessed June 25, 2020, https://irbo.nih.gov/confluence/.

U.S. Department of Health and Human Services. 1996. "Informed Consent Requirements in Emergency Research (OPRR Letter)." Accessed June 25, 2020, https://www.hhs.gov/ohrp/regulations-and-policy/guidance/emergency-research-informed-consent-requirements/index.html .

CHAPTER 

Disseminating and Archiving Results

LEARNING OBJECTIVES

Upon completion of this chapter, you should be able to:

- Describe the various ways to disseminate research.
- Discuss how to archive your results.

Case Study

The Somewhereville Fire Department conducted an improvement project that significantly reduced the time from arrival at the scene to the first water on the fire.

1. Where might this project be submitted for publication?
2. What conference(s) might you consider for presenting the project?
3. What is the process for getting your project accepted for presentation at one of those conferences?

Introduction

Once the research or process improvement project has produced results, the project should not be considered over until the information has been shared and archived. Externally, there are others who may face similar challenges or have similar questions for which they are seeking answers. They will want to learn from your project. Internally, all of the people who have been involved in the project will want to know what information and what benefit their work has yielded.

Chances are, there will be similar challenges and questions at some point in the future. To avoid having to repeat much of the same work when those occasions arise, archiving the raw data, notes, analyses, and documents from your project in appropriate formats will be an invaluable resource.

Project Presentations

The appropriate format for the presentation of the results of a research or process-improvement project depends on the setting and the audience. Quite often, the presentation needs to be in both written and oral formats.

Written Formats

The written format allows an internal or external audience to take more time to understand the findings and results. Written presentations usually fall into one of three general categories: abstracts, executive summaries, and full technical reports.

Written Abstracts

An **abstract** is a very concise and structured summary of a project. In this era of information overload, abstracts are a useful tool for readers to get the essence of a project and decide if they need to read the full technical report. Therefore, the abstract may be the only output of a project that many, if not most, will ever read. That is why consideration should be given to writing the abstract as clearly and succinctly as possible. Professional associations and publishers that accept abstracts as part of a submission for presentation or publication often require a very specific abstract format.

Commonly, an abstract is limited to 250 words or less. The format usually includes the following components:

- Introduction. The introduction should identify the problem, issue, subject, or reason why the project was conducted. The setting where the project was conducted is briefly described. The hypothesis, if one is used, is stated here as well.
- Methods. The process used for conducting the project is described. This will often include how data were collected and analyzed.
- Results. The highlights of the findings are outlined. For quantitative studies, this may include a few key statistics. The results of the hypothesis test, if applicable, are included.
- Conclusion. This section should convey the significance of the project and its key findings.

Given the need to communicate so much in so few words, the abstract can be one of the most challenging parts of writing up a project. Many find writing an abstract before writing a full technical report to be a very useful exercise because it forces the author to clarify the message.

Written Executive Summaries

Some internal reports and external publication formats include an **executive summary** as a separate section before the full technical report. As with the abstract, many readers may not have the time or inclination to read the full technical report. The executive summary covers the highlights and key conclusions. The format is not as structured as an abstract and usually has much more flexibility in its length. It should generally not exceed a page or two in most circumstances. With abstracts and executive summaries, less is more.

Full Technical Reports

A **full technical report** may be in the form of a detailed scientific paper for publication or a detailed internal document. The length will vary with the nature of the work, the audience, and where the report is published or distributed. If the report is intended for publication in a journal, there will usually be constraints on the length. Internal reports will have more latitude. In either case, authors should exercise discipline to communicate concisely. Resist the temptation to write anything and everything that comes to mind. Good technical writing respects the time of the reader.

The format of a full technical report often follows the same general format as a structured abstract. Some professional societies and publishers prescribe specific formats, just as they often do for the abstract.

A full technical report may provide enough detail for readers to replicate the study, if so desired. The objective is not necessarily to have someone actually replicate the study but to allow them to have enough of an understanding of how the study was conducted and how the data were analyzed to determine whether the problem, methods, and conclusions fit together appropriately. The details of the methods and analysis should allow the reader to judge whether the processes used were appropriate to the problem and were conducted in a technically sound manner.

When the full technical report is published on its own—not in a magazine or journal—the author may include appendixes. Free-standing documents include "white papers," monographs, books, and internally distributed reports. The appendixes in a full technical report may include highly detailed descriptions of certain procedures and copies of data-collection forms. They may also include tables of raw data if these may be useful to the reader. By putting these sorts of items in appendixes, readers have the option of seeing more details if they want to.

Poster Presentations

Poster presentations are usually made at professional conferences. After submitting a written abstract, conference organizers may invite the authors to present their work in an oral abstract session or as a poster presentation.

Poster presentations, which may include both text and graphics, are usually put on large, double-sided poster easels. This allows separate poster displays on each side of the easel. Rows of these easels are often set up in a designated area in the conference venue (**FIGURE 10-1**).

A typical poster display area is 4 feet (1.2 m) wide and 3 feet (0.9 m) tall. The poster area can be filled with smaller sheets of 8.5 in. (21.5 cm) by 11 in. (30 cm) or larger sheets tacked into place on the poster display easel. Colored paper backgrounds are often used to frame each sheet to make it look more professional. With today's widespread availability of wide-format printers at copy centers, many project leaders now choose to compose an image that is 4 foot (1.2 m) by 3 foot (0.9 m) and includes blocks of text and graphics on a background that prints at that size—a true poster (**FIGURE 10-2**).

In a poster display, authors usually have some latitude to expand on the words in their written abstract and are encouraged to include graphics. At the meeting, attendees roam through the poster displays, reading at their leisure throughout the time of the conference. Some conferences schedule a specific time for authors to stand at their posters so that attendees can come by to discuss the project and ask questions.

Articles

Articles are associated with some sort of publication. They usually have a less technical tone and are shorter in length. Much depends on the guidelines that are specific to each publication. Trade publications may take a much less formal tone and include more photographs and illustrations to break up blocks of text for a more visually appealing page style, which may often be mixed with advertisements. Some organizations

FIGURE 10-1 Poster presentations at conferences are ways to disseminate your research findings.

FIGURE 10-2 An example of a poster presentation.

may have magazines or newsletters primarily for internal distribution, in which projects of interest to the entire workforce may be presented.

Oral Formats

Oral formats allow audiences to get an understanding of the problem or topic along with the findings and results—particularly when they have not had time to read a written version. An oral presentation also allows the audience to discuss and ask questions. Oral formats usually fall into the general categories of abstracts, technical presentations, and lectures (**FIGURE 10-3**).

Oral Abstracts

Oral abstracts are usually associated with presentations of formal research at scientific meetings. They follow the same general format as a written abstract, but instead of a word-count constraint, there is a time constraint. Commonly, an oral abstract presentation is limited to 10 minutes, plus 5–10 minutes for questions from the audience. Presentations are typically made with the visual aid of electronic slides (e.g., PowerPoint or Keynote). Usually, the audience includes other researchers and technically savvy professionals. It is accepted decorum for audience members to ask difficult questions that challenge the methods, data analysis, and conclusions of the presenters. This is a form of peer review, which will be discussed in more detail later in this chapter.

Oral Technical Presentations

Like their written counterparts, oral technical presentations can drill into as much detail as the venue, audience, and available time permit. These types of presentation are most commonly made for internal audiences that have a specific interest in the project as it relates to their own responsibilities in their organization. For example, a technical presentation on a research project or process improvement project to improve a specific aspect of fire-ground operations would likely be made to fellow officers in the same fire department. These presentations often use electronic presentation software as a visual aid. Again, avoid the temptation to ramble and say everything that comes to mind. Respect the time and attention of the audience. Be concise, but also be ready to go into more detail in response to questions from the audience.

FIGURE 10-3 Oral formats presented to a group of individuals.

Courtesy of Estero Fire Rescue.

Lectures

Lectures are usually given at trade or professional association conferences. They are typically around 45–60 minutes in length and can have an audience of anywhere from a dozen to thousands. Electronic slides are often used as a visual aid, but this is not required. Many extremely effective lecturers do not use electronic slides, or use them sparingly. Too often, lecturers rely too much on electronic slides. Used inappropriately, electronic slides can detract from audience interaction and dilute the overall quality of the lecture (Kerr 2001; Reynolds 2008). Time is usually set aside near the end of the allotted lecture time for discussion and questions from the audience.

Peer Review

When a project is submitted to a professional association for publication in an academic, scientific, or technical journal, it will undergo the **peer-review process**. Given the limited number of pages in a journal issue and the limited amount of time at a scientific conference, the peer-review process selects those projects that have the most importance and technical merit. Even if more space or time is available, peer review seeks to reject projects that do not meet scientific standards, particularly in terms of appropriate project methods, data analysis, hypothesis testing, interpretation, and logical support of conclusions made.

For a publication, peer review is usually done by members of an editorial board and other technical subject-matter experts who serve as reviewers and supplement the input of editorial board members. The editor of the publication will complete a preliminary review and select several editorial board members and reviewers to scrutinize the paper. Reviewers are selected on the basis of their individual expertise in the topic covered by the paper. The subject-matter experts may be complemented by a statistical expert who can address the technical aspects of more complex types of analyses. These individuals will usually score the paper on a scale of 1–5 on criteria such as the following:

- Relevance to the publication's audience (For a journal on fire operations, is this an article in which fire operations chiefs would be interested?)
- Scientific/technical quality (Were appropriate methods used to conduct the project and analyze the data?)
- Quality of writing (Was the information expressed clearly and concisely?)

There is also an opportunity for reviewers to offer positive comments and constructive criticisms that may help improve the paper.

The peer reviewers are usually asked to explicitly recommend one of the following actions to the editor:

- Accept for publication in its current form.
- Accept for publication contingent on making specific additions or corrections.
- Reject in current form, but invite resubmission after major issues are addressed.
- Reject without invitation for resubmission.

The comments of reviewers and the scores are then shared with the authors, along with the decision of the editor, who makes the final call based on the combination of input from all reviewers and his or her own assessment. Some comments from peer reviewers are intended only for the editor and are not shared with the authors.

In open peer review, the peer reviewers are told the identity of the authors, and the authors are told who the reviewers are. Because this can lead to some types of bias, closed peer review, in which peer reviewer and author identities are not disclosed to each other, is more common.

Because of the quality control that peer review provides, papers that are published in journals that use a formal peer-review process are often considered to be much more credible than others.

Archiving a Project

A problem or issue appeared, and a project was begun to address the problem. Maybe the project began as a way to make something that already worked work even better. It turned into a proposal. Management gave the OK. A project team was formed. Data collection and analysis followed. Conclusions were reached. Lessons were learned. Oral and written reports were made. Is the project over? Almost.

A significant amount of time can be spent even on a comparatively small project, considering where it starts at the problem identification or process-improvement idea stage. Along the way, a very significant amount of information can be generated—emails, phone calls, memos, proposals, raw data, statistics, graphs, reports, and maybe more.

Should the solution that was implemented to fix the problem or improve the process backslide for whatever reason, there can be huge value in having access to all the information that was previously generated. The amount of time and money this can save can be huge—it won't be necessary to repeat the same meetings or collect the same data or even try to remember why specific conclusions and recommendations were made.

The problem with most projects is that very little of the information generated by the project is saved afterward. There may be a couple of paragraphs outlining the new policy or procedure, and that's it. Sometimes, you are very, very lucky if some sort of report is produced and saved.

Archiving is a logical strategy to adopt if you think proactively about a scenario in which backsliding happens. Access to the original project information will help get things back on track. In mainstream business, this is part of a larger theme that is often called *organizational learning and knowledge management*. Archiving is a key component of how an organization, such as a fire department, learns and shares the collective data, experiences, and insights gained by all of its individual members.

Storing all the paper files—notes, reports, and so forth—in a file cabinet is better than nothing, but it is far from optimal. Often, the file is never looked at again because someone at some point in the future does not know about the file or what it contains. That is where electronic files have their strength. They can be searched electronically using many different types of computer utilities, ranging from simple search tools built into computer operating systems to commercial knowledge management software systems.

Regardless of what tools you use for searching, at the end of each project, make the final step a process in which all of the information associated with the project is put into electronic form. All of those electronic files should be stored in a specific folder. That folder should be put into a specific hard drive on your organization's data system. It should also be backed up onto physical media (e.g., CDs or DVDs). A copy of that media should be stored in another off-site location in case of any type of disaster (fire, flood, storm, theft, etc.) at your offices. When the time comes, these electronic resources will be worth their weight in gold.

Wrap-Up

CHAPTER SUMMARY

Having gone through all that effort to conduct your research, you want to be able to share the knowledge you gained with others. In this chapter, we have discussed a variety of ways to do this. The body of knowledge in the fire service cannot be used for improvements elsewhere if you don't share it. What you may learn from experiences in your research may be a source of knowledge that will prevent an event or help someone just like you in another fire department. Share the knowledge and read the literature that has already been published on past research projects.

KEY TERMS

Abstract A very concise and structured summary of a project.

Executive summary Covers the highlights and key conclusions of a project.

Full technical report A detailed scientific paper for publication or a detailed internal document.

Peer-review process A process that selects those projects that have the most importance and technical merit.

Poster presentations Can include both text and graphics and are usually mounted on large, double-sided poster easels.

REVIEW QUESTIONS

1. Compare and contrast the three different written formats for the presentation of results.
2. Describe the various oral presentations for disseminating research findings.
3. You are going to submit your research findings to a journal that requires peer review. Explain what this means.
4. Describe the components of an abstract.
5. What does archiving consist of?
6. Describe the various ways to disseminate research.
7. What are the peer reviewers usually explicitly asked to recommend to the editor?
8. What are the benefits and disadvantages of presenting the information in a lecture format?
9. Your research project has been accepted for a poster presentation at a national conference. Describe what you need to do to prepare and what will likely be expected of you at the conference.
10. What are the various means by which technical reports are published?

REFERENCES

Kerr, C. 2001. *Death by PowerPoint: How to Avoid Killing Your Presentation and Sucking the Life Out of Your Audience, Your Effective Tip-Kit for the Effective Use of PowerPoint.* Santa Anna, CA: ExecuProv Press.

Reynolds, G. 2008. *Presentation Zen: Simple Ideas on Presentation Design and Delivery.* Berkley, CA: New Riders.

CHAPTER 11

Sources for Research on Fire-Related Activities

LEARNING OBJECTIVES

Upon completion of this chapter, you should be able to:

- Describe the various organizations and other resources that conduct research on fire-related activities.

Introduction

Various sources are available that can help you identify and review research on fire-related subjects. This chapter looks at the various agencies that conduct fire-related research or that can provide links to fire-related research. The sources vary in the depth of research they conduct. Each source contributes to the intricacies of research in the fire service according to its mission. There is a wide spectrum of fire-related research, and the sources cited in this chapter cover a diverse sampling of the agencies that conduct research.

National Institute of Standards and Technology

The **National Institute of Standards and Technology (NIST)** develops technologies, measurement methods, and standards. NIST works with industry to overcome technological hurdles that limit quality and innovation. For several years, NIST has worked with the Interagency Board for Equipment Standardization and Interoperability (IAB), the National Fire Protection Association, the National Institute of Occupational Safety and Health, the Edgewood Chemical Biological Center, and other partners to develop the first performance and compliance standards for hazardous materials (hazmat) suits and other equipment designed to protect first responders from chemical, biological, radiological, and nuclear threats (**FIGURE 11-1**). These standards should give emergency service personnel confidence that their equipment will provide adequate protection and function reliably.

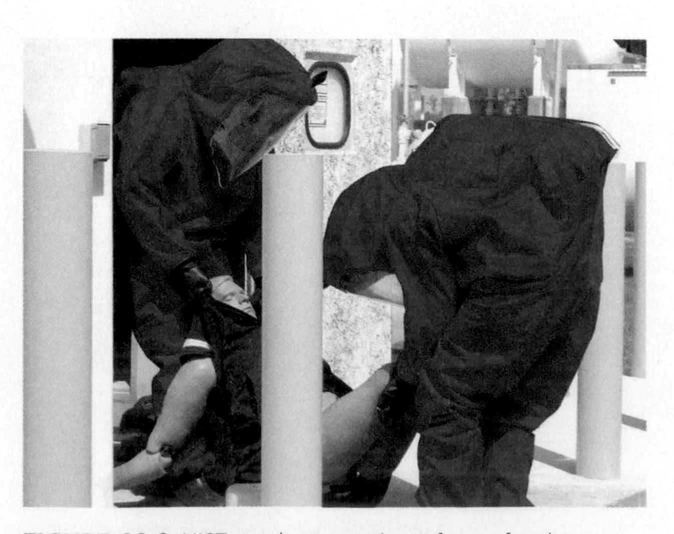

FIGURE 11-1 NIST conducts a variety of tests for the fire service. Part of the testing is on personal protective equipment, such as hazardous materials suits.

Courtesy of Estero Fire Rescue.

NIST conducted a 2-year, $16 million investigation of the structural failure and collapse of the World Trade Center buildings following the attacks of September 11, 2001. The investigation was intended to determine what happened and why. Conclusions from the study were used to recommend changes to building and fire codes, standards, and practices, as well as to improve emergency response and evacuation procedures.

A team of NIST scientists continues to work to help emergency personnel save more people who may be trapped in collapsed buildings. Before the demolition of a high-rise building in New Orleans, the NIST team placed specially modified radio transmitter modules, operating in the frequency bands used by emergency personnel and mobile telephones, at various points throughout the structure. The researchers collected information on the transmitters' signal strength and other data. The project's goal was to foster the development of technologies that allow emergency personnel to lock on to cell phone or radio signals within collapsed buildings so as to locate and communicate with first responders and survivors.

NIST-designed standard test courses are helping to improve "rescue robots" able to navigate spaces in collapsed buildings that are too small or too hazardous for people to access. Test arenas patterned after the NIST design have been built in the United States, Japan, and Italy and are expected soon in Germany and Portugal. Team competitions held at the arenas are advancing rescue robot technologies.

Four technical programs at NIST focus on core research that affects the fire service:

- High-Performance Construction Materials and Systems. This program enables scientific and technology-based innovation to modernize and enhance the performance of construction materials and systems.
- Fire Protection Technologies. This program enables engineered fire safety for people, products, and facilities and enhanced fire-fighter effectiveness with a 50 percent reduction in fatalities.
- Enhanced Building Performance. This program provides the means to ensure that buildings work better throughout their useful lives.
- Homeland Security. This program uses the lessons learned from the World Trade Center disaster to better protect people and property, enhance the safety of fire and emergency responders, and restore public confidence in the safety of tall buildings nationwide.

The research that NIST conducts is at a higher level than most research projects. The funding and complexities make it difficult to emulate this level of research at the fire department level. Fire agencies may be able to participate in the research that NIST conducts rather than attempting to conduct research at this level.

National Fire Protection Association

The **National Fire Protection Association (NFPA)** is well known in the fire service. The association is a member-based organization. The NFPA's wealth of fire-related research includes investigations of technically significant fire incidents, fire data analysis, and the Charles S. Morgan Technical Library, one of the most comprehensive fire literature collections in the world. In addition, the NFPA's Fire Protection Research Foundation is a source of independent fire test data (Almand 2008).

The NFPA Fire Protection Research Foundation has been given the mission to plan, manage, and communicate research in support of the NFPA mission. The focus of the research has included both international and domestic issues. The research is designed to provide the type of information that the NFPA's technical committee members and others can use to better support fire safety codes and standards. Each project is guided by a project technical panel that provides technical expertise and user input from sponsors, the research community, the fire service, NFPA technical committees, and other stakeholders.

A number of NFPA research projects have significantly affected the fire service. In the 1980s, the Fire Protection Research Foundation and the NFPA worked together to find alternatives to Halon. Halon was a chemical component that was used as an extinguishing agent, especially in the computer industry, but Halon was considered to have negative effects on the Earth's protective stratospheric ozone layer. Through symposia and a report on the wasteful full-scale field testing of Halon 1301 fire-suppression systems, the Fire Protection Research Foundation and the NFPA were able to help transition to more environmentally friendly fire protection chemicals. The report on this subject, published in 1988, was *Halon 1301 Discharge Testing: A Technical Analysis.* Through more efforts and reports, the final result led to NFPA 2001, *Standard on Clean Agent Fire Extinguishing Systems*. The process used during this change still serves as a classic example of the way research results are used in the NFPA's development process for codes and standards.

The NFPA and the Fire Protection Research Foundation have also worked with other agencies on research projects. One example is fire risk assessment. During the latter half of the 1980s, the Fire Protection Research Foundation sponsored a joint project with NIST, the NFPA's Fire Analysis and Research Division, and Benjamin-Clarke Associates to develop a generic fire risk assessment package for evaluating alternative products. The motivation for the project was to move away from the then-active push to evaluate products solely on the basis of their toxic-combustion potency.

The model developed during the project, which would be named the Fire Risk Assessment Method, or FRAMEworks, combined a fire scenario and a behavioral scenario structure with associated probabilities, a zone model for fire development and smoke spread, a smoke alarm activation model, and a toxic-impact model.

The second project developed a fire risk assessment framework to evaluate fire protection alternatives for "big-box" retail and storage facilities in which large quantities of flammable or combustible liquids are stored in consumer-ready containers. Sprinklers of various designs can be compared to innovative packaging design, floor drains, and other strategic approaches.

The Fire Protection Research Foundation has also made many significant contributions to the understanding of fire sprinkler technology and fire detection and alarm systems. The Fire Protection Research Foundation provides an independent resource for the "code road mapping" of new fire technology as well as an effective means of identifying the performance criteria against which such technology will be evaluated.

Today, the Fire Protection Research Foundation is responding to current challenges with activities in a number of areas, including detection and signaling, hazardous materials, electrical safety, fire suppression, storage of commodities, and fire-fighter protective clothing and equipment. You can find the reports by the Fire Protection Research Foundation online at its website.

U.S. Fire Administration

U.S. Fire Administration (USFA) programs prevent and mitigate the consequences of fire. These programs fall into four basic areas:

Public Education. Develops and delivers fire prevention and safety education programs in partnership with other federal agencies, the fire and emergency response community, the media, and safety interest groups.

Training. Promotes the professional development of the fire and the emergency response community and its allied professionals. To supplement and support state and local fire service training programs, the National Fire Academy

(NFA) develops and delivers educational and training courses that have a national focus.

Technology. Works with the public and private groups to promote and improve fire prevention and life safety through research, testing, and evaluation. Generates and distributes research and special studies on fire detection, suppression and notification systems, and fire and emergency responder health and safety.

Data. Assists state and local entities in collecting, analyzing, and disseminating data on the occurrence, control, and consequences of all types of fires. The National Fire Data Center describes the nation's fire problem, proposes possible solutions and national priorities, monitors resulting programs, and provides information to the public and fire organizations (U.S. Fire Administration n.d.).

National Fire Data Center

The mission of the **National Fire Data Center** is the collection, analysis, publication, dissemination, and marketing of information related to the nation's fire problem and USFA programs (**FIGURE 11-2**). The center manages USFA research efforts in fire detection, prevention, and suppression and first responder health, safety, and effectiveness (U.S. Fire Administration n.d.).

The functions of the National Fire Data Center are to:

1. Coordinate and effect the collection, analysis, and dissemination of information about fire and other emergency incidents involving fire department response.
2. Manage the National Fire Incident Reporting System, Hotel-Motel National Master List, and other national databases containing fire and hazardous materials information.
3. Coordinate fire service issues with other Federal Emergency Management Agency (FEMA) directorates and offices and with other government agencies, national fire organizations, and allied professions.
4. Review and authorize reimbursement to local fire services for firefighting on federal property.
5. Administer the Learning Resource Center, which provides library and information services to students, staff, and faculty at the National Emergency Training Center (NETC), and provide reference services to the nation's fire and emergency management communities.
6. Encourage and assist state, local, and other agencies, public and private, in developing standardized reporting methods and reporting information.
7. Create, maintain, and disseminate—to the public, students, staff, and faculty at the NETC—operational, technical, educational, and marketing information in a variety of formats through the administration of the USFA Publications Center, USFA World Wide Web site, and USFA Media Center.
8. Manage the research, development, and application of projects and investigation of technology for fire detection, prevention, rescue, and suppression, as well as technology, equipment, and strategies to improve fire-fighter and first responder health, safety, and effectiveness.
9. Plan and coordinate USFA participation in conferences, shows, and exhibits.

FIGURE 11-2 The National Fire Data Center is located on the campus of the National Fire Academy. The National Fire Data Center manages USFA research efforts in fire detection, prevention, and suppression and first responder health, safety, and effectiveness.

Underwriters Laboratories

Underwriters Laboratories (UL) has more than 114 years as one of the world's leading product safety testing and certification organizations. In the fire service, the UL Mark is one of the most recognized symbols of safety in the world (Underwriters Laboratories n.d.).

UL is an architect of U.S. and Canada safety systems, having developed more than 1200 safety standards and participating actively in national and international standards development. UL tests more than 19,000 types of products, and 21 billion UL Marks appear in the marketplace each year (**FIGURE 11-3**).

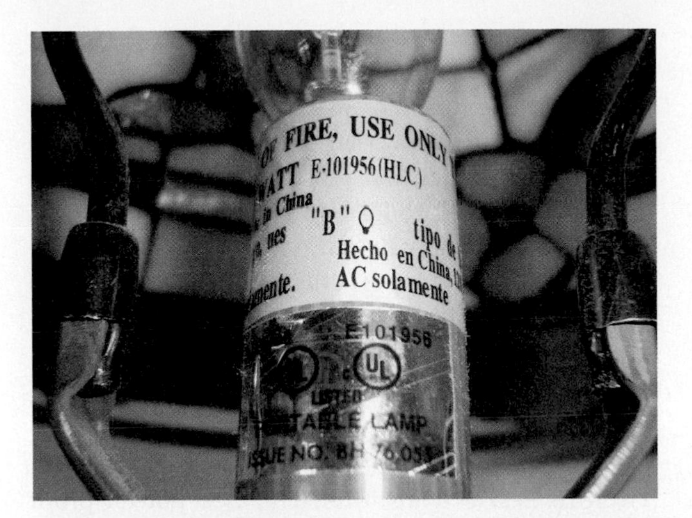

FIGURE 11-3 The UL Mark appears on many products.

Consumers and regulatory authorities value the UL as a leader on safety issues. With public safety at the heart of the UL's mission, the UL acts as a safety resource and advocate. The UL works closely with customers, regulators, insurers, retailers, and consumers on research, technology, and safety initiatives. UL also promotes public safety through education and outreach through and to the media.

Working with more than 72,000 customers in 99 countries, the UL helps enhance safety and quality on a global scale. With 63 laboratory, testing, and certification facilities worldwide, the UL is the standard in safety.

The U.S. Department of Homeland Security (DHS) recently awarded the UL a $991,900 Fire Prevention and Safety Research Grant to enhance understanding of the hazards to fire fighters in structural fires and provide data to further advance knowledge of current firefighting tactics.

Conducted in cooperation with the International Association of Fire Chiefs (IAFC), the Chicago Fire Department, and the University of Maryland Fire Protection Department, the Firefighter Safety Research Project will (1) investigate the structural stability of engineered lumber and (2) evaluate the effectiveness of extinguishing agents used to fight fires in modern structures.

Earlier research by the National Engineered Lightweight Construction Fire Research Project indicated that unprotected lightweight wood truss assemblies can fail within 6 to 13 minutes of exposure to fire. Between 1998 and 2003, the National Institute of Occupational Safety and Health attributed 13 firefighter fatalities and 9 firefighter injuries to the collapse of buildings built with lightweight wood trusses. During the same time frame, five fatalities and two injuries were attributed to collapses of buildings with heavy-timber, solid-joist lumber truss construction.

The second part of the research project will evaluate the effectiveness of various extinguishing foams in residential fire situations and provide information to help national fire service organizations design new firefighting tactics, develop web-based training programs, and increase overall fire-fighter safety.

The Firefighter Safety Research Institute is the UL's latest research and development project to advance the DHS goal to eliminate residential fire deaths by 2020. The UL completed a ground-breaking study that investigated 27 synthetic and natural materials and various combinations of materials now most commonly found in homes. As a result, the UL is now strongly recommending that consumers utilize both photoelectric and ionization technologies to optimize detection and permit the best available escape time in residential fire situations.

Fire Research and Management Exchange System

The **Fire Research and Management Exchange System (FRAMES)** is a systematic method of exchanging information and transferring technology among wildland fire researchers, managers, and other stakeholders (**FIGURE 11-4**). In partnership with the U.S. Geological Survey's National Biological Information Infrastructure (NBII) Program, FRAMES implements web-based technologies that can help bridge the gap between science and management. The goal is to make wildland fire data, metadata, tools, and other information resources easy to find, access, distribute, compare, and use. In collaboration, the wildland fire research and management communities can use these technologies

FIGURE 11-4 FRAMES focuses its research on wildland fires and firefighting.

to help eliminate redundancy, reduce costs, and promote increased productivity and efficiency (Fire Research and Management Exchange Systems n.d.).

FRAMES is not a website that merely points to other websites. Although indexes of wildland fire websites may be useful starting points, they are often frustrating when you are looking for specific content. Content management at FRAMES is more integrative and gives users more options.

There are currently three receptacles for content in FRAMES: Resources, Services, and Library. From a file management perspective, you can think of each of these as folders that contain subfolders with files in them.

Resources. This folder contains a dynamic stock of assets that can be useful for solving wildland fire problems. They are sources of information or functionality that may be of particular utility to FRAMES users. Currently, Resources contains three subfolders: Data, Documents, and Tools. Data are information resulting from measurement, analysis, or modeling structured for subsequent analysis. Documents are narrative sources of information, such as articles, books, and correspondence. Tools are devices that modify or transform input in a significant way to produce output. Each of these subfolders contains files, which are metadata records for the individual content item, such as data, a document, or a tool. To be added to the Resources folder at a later date are the subfolders Projects and Programs. These will likewise contain metadata records about individual projects and programs.

Services. This folder is currently being developed. When it is implemented, it will contain utilities to aid, assist, or facilitate the wildland fire research and management communities. Services are likely to include subfolders such as Threaded Discussions, Project Management, News, and Notices.

Library. This folder contains a collection of reference material, including an encyclopedia, glossaries, images, thesauri, user guides, and offsite links.

Fire.gov

In November 1997, NIST and the USFA announced plans to collaborate and work to bring firefighting to new levels via ongoing research activities around the world. The website contains contact information and links to specific researchers and organizations involved in the research. So often, it is difficult to locate research that has been conducted. This website helps the researcher find those previously conducted research reports. The Fire.gov website has expanded to provide training materials, videos, fire reconstructions, and research reports that may be of interest to the fire service and the fire researcher (National Institute of Standards and Technology n.d.).

Fire Research Institute

The **Fire Research Institute (FRI)** is a not-for-profit library holding over 80,000 books, journal articles, videos, training manuals, dissertations, news reports, and other material on wildland fires. The FRI provides a free monthly electronic newsletter listing new publications that have been released (Fire Research Institute n.d.).

Fire Research Laboratory

The **Fire Research Laboratory (FRL)** is an innovative partnership among law enforcement, fire services, public safety agencies, academia, and the private sector that uses the most advanced scientific, technical, educational, and training methods to make the U.S. Bureau of Alcohol, Tobacco, Firearms, and Explosives (ATF) and its partners leaders in fire investigation science to serve and protect the public.

The strategic goals of the FRL are as follows:

- Conduct scientific research that validates fire scene indicators and improves fire scene reconstruction and fire evidence analysis.
- Support fire investigations and the resolution of fire-related crimes.
- Develop improved investigative and prosecution procedures using scientifically validated methods that integrate the assets of ATF and its partners to enhance fire investigation personnel expertise.
- Establish a central repository for fire investigative research data that will be disseminated throughout the fire investigation community.
- Develop an internationally recognized research and education center for the advancement of knowledge, technology transfer, and case support related to fire cause investigation and fire scene reconstruction.

The initial concept of an FRL developed out of a demonstrated need by ATF Certified Fire Investigators

(CFIs) to assist them in the process of fire scene reconstruction to identify scientific-based theories for fire ignition and development (ATF Fire Research Laboratory n.d.). Consequently, Congress directed the ATF to establish a research laboratory dedicated to advancing the science of fire investigation.

In the fall of 1997, ATF convened an International Conference on Fire Research for Fire Investigation with the assistance of the University of Maryland's Department of Fire Protection Engineering, the NIST Building and Fire Research Laboratory (BFRL), and Hughes Associates, Inc., a private fire protection engineering consulting firm based in Baltimore, Maryland. The conference was attended by approximately 70 leading authorities from the fields of fire investigation, fire science, fire research, training, and education from the United States and five foreign countries for the following purposes:

- To assess the current state of the art of fire investigation and its use of scientific principles and methodologies
- To identify fire investigation needs for research and education
- To recommend the role that the FRL should play in advancing fire investigation and research
- To recommend capabilities and staffing for the FRL to accomplish its mission successfully

The input received from these resources formed the blueprint for the new laboratory.

The laboratory's technical planning and development is a partnership among the ATF/New Building Projects Office (NBPO), Factory Mutual Research Corporation (FMRC), the University of Maryland's Department of Fire Protection Engineering, and NIST's BFRL.

The FRL undertakes research, education, and case support for fire investigation and analysis on behalf of ATF CFIs, prosecutors, and the fire investigation community. The FRL provides the opportunity and facilities for scientists, engineers, and researchers to work on important fire investigation issues, such as the following:

- Scientific research directed at the determination of fire origin and cause, fire growth and spread, and fire scene reconstruction that validates fire scene indicators and improves fire evidence analysis
- Case support for ATF fire investigations and the conviction of arsonists
- Development of improved investigation and prosecution procedures based on scientifically validated methods that integrate the assets of ATF and its partners to enhance fire investigation personnel knowledge and expertise
- Establishment of a central repository of fire investigation test data that will be accessible to the fire investigation community to improve public safety
- Development of an internationally recognized research and education center for the advancement of knowledge, technology transfer, and case support related to fire cause investigation and fire scene reconstruction

Fire investigation testing and training are conducted for ATF CFIs and other public safety organizations with fire investigation responsibilities. The laboratory provides a controlled environment where fire investigation theories can be evaluated and fire cause determination scenarios can be reconstructed and tested on a large scale. FRL staff are able to conduct demonstrations of fire phenomena for training purposes and conduct unique research in fire behavior as it relates to fire origin and cause determination. The FRL staff provides training that produces a cadre of accredited investigators and CFIs.

The FRL is an active participant in the worldwide community of fire research laboratories and serves as an international model in training personnel, developing research and testing protocols, and fostering technical partnerships that helps the ATF maintain credibility in the broader fire research and testing community.

The primary equipment used in these laboratories is hood/exhaust systems that allow for the measurement of the heat release rates (HRRs) of burning materials. The laboratory provides a controlled environment where fire scientists, engineers, and researchers perform a wide range of tests to evaluate fire investigation theories and reconstruct and test fire cause determination scenarios on a large scale.

The FRL has several large test bays (cells) where test fires are conducted. The largest fire test cell is approximately 130 ft (39.6 m) by 130 ft (36.9 m) by 55 ft (16.7 m) in height and has the capability to accommodate multiple-room, vehicle, and two-story-structure fire scenarios. These facilities provide the ATF with the versatility it needs to reconstruct and test key aspects of most of the fire scenarios encountered by fire investigators in the field.

In addition to the large test bays, the laboratory has reconfigurable small-scale test areas and bench-scale test equipment to predict fire behavior prior to actual large-scale fire tests being conducted. Additional support space in the FRL includes the following:

- An electrical testing laboratory space
- A state-of-the-art fire control test center and fire safety suppression system with associated on-site air and water pollution treatment facilities
- Observation space for visitors and observers
- On-site classroom and training space for approximately 50 persons
- Support spaces that include shop areas, showers, material conditioning rooms, construction/test materials storage, and evidence storage areas

The FRL is staffed by a number of scientists, engineers, and researchers from the disciplines of fire protection engineering; mechanical, chemical, materials, and electrical engineering; and metallurgy, as well as ATF CFIs. The FRL also provides opportunities for undergraduate and graduate students to work in the laboratory to assist FRL scientists and engineers with fire research and testing projects while completing their academic requirements.

The FRL joins two existing laboratories that occupy ATF's National Laboratory Center (NLC) complex built on a 35-acre site in Beltsville, Maryland, approximately 8 miles north of the Washington Beltway. The NLC is the administrative headquarters for the ATF Laboratory System that encompasses three forensic laboratories (Washington, DC; San Francisco, California; and Atlanta, Georgia). The NLC complex houses the ATF Alcohol and Tobacco Laboratory (ATL), the Forensic Science Laboratory—Washington (FSL-W), and the new FRL. The FRL occupies approximately 50,000 gross square feet of the new facility and is operated by the ATF Laboratory Services Division, Office of Science and Technology.

The FRL serves as a central repository of scientific information related to fire incident investigation, analysis, and reconstruction research. The FRL disseminates the knowledge derived from this research through scholarly publications in the scientific and investigation literature, through training and education programs, and via electronic access.

Fire investigation training programs are also conducted for ATF CFIs and other public safety organizations with fire investigation responsibilities. FRL staff members are also able to use the laboratory facilities to conduct "live" demonstrations of fire phenomena for training purposes.

Institution of Fire Engineers

The **Institution of Fire Engineers** was founded in 1918 in the United Kingdom by a group of eight British chief fire officers (**FIGURE 11-5**). The mission statement they adopted then continues to guide the institution in meeting the needs of its members and serving the interests of society:

> To promote, encourage, and improve the science and practice of Fire Engineering, Fire Prevention, and Fire Extinction, and all operations and expedients connected therewith, and to give impetus to ideas likely to be useful in connection with or in relation to such science and practice to the members of the Institution and to the community at-large. (The Institution of Fire Engineers n.d.)

The U.S. branch continues the tradition of fire service leadership in promoting fire engineering established by the forward-looking men who established the original institution decades ago and an ocean away. A similarly small group of fire service leaders met in Tulsa, Oklahoma, in February 1996 to establish the branch.

- The institution has almost 60 branches worldwide that run seminars and 1- to 3-day conferences and have their own newsletters.
- The institution is forming special-interest groups with a nongeographic basis.
- The institution has over 11,000 members worldwide, occupying positions of responsibility and influence in all aspects of the fire world. All members are awarded appropriate certificates to validate their status.
- The institution conducts examinations of its members in all countries where it has branches and even in those countries where there is a requirement but no established branch.
- The institution provides and administers the forum for educational qualifications in the fire community, involving a network of universities and colleagues throughout the country, and actively supports the Chair of Fire Engineering at the University of Central Lancashire.
- The institution accredits and validates courses of study and training for numerous outside educational bodies and training establishments.
- The institution secures and administers bursaries, educational grants, and scholarships on behalf of the Fire Service Research and Training Trust and for other organizations.
- The institution actively encourages education and training in underdeveloped countries through its Traveling School.

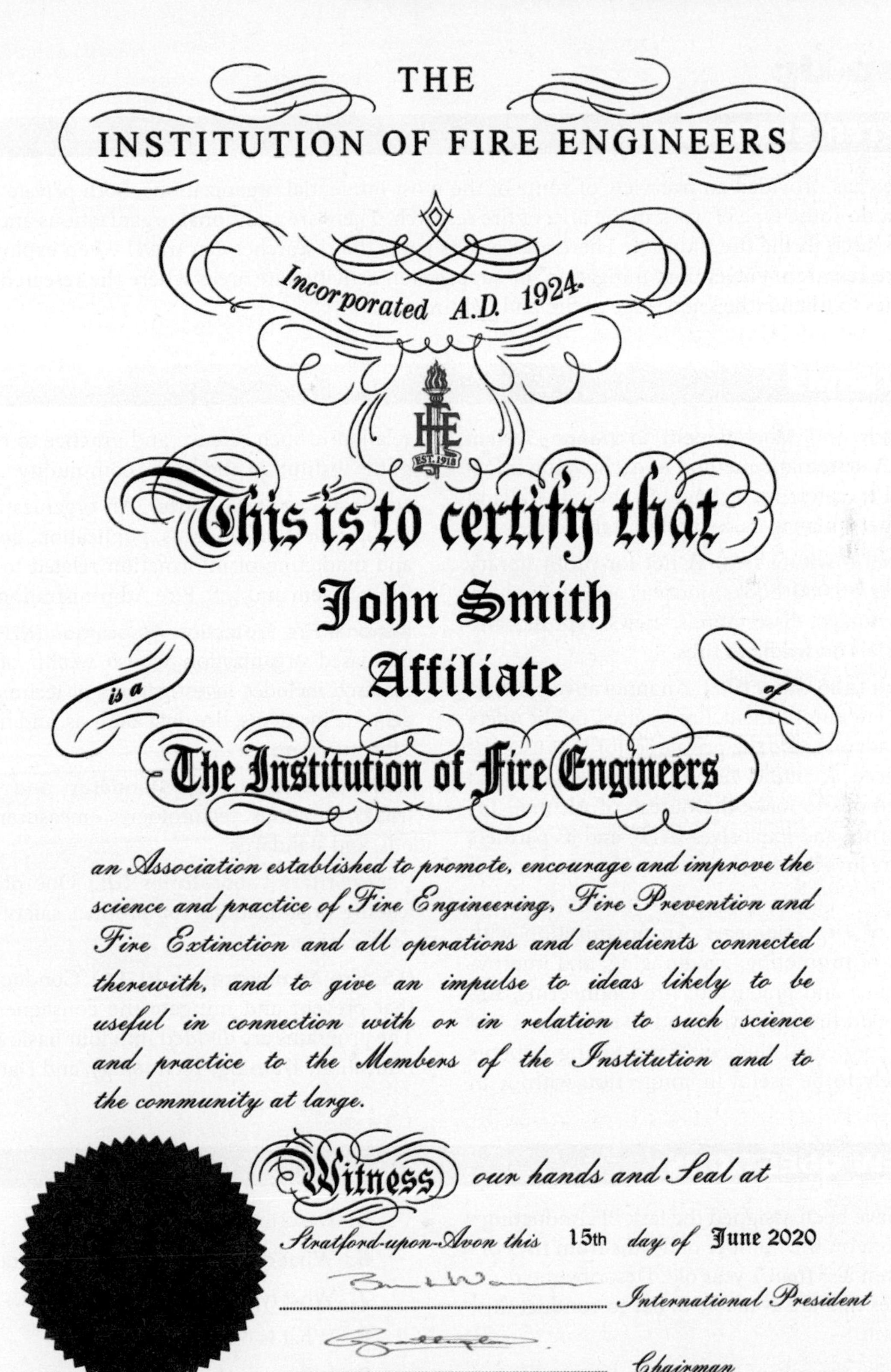
THE
INSTITUTION OF FIRE ENGINEERS
Incorporated A.D. 1924.
IFE
EST. 1918
This is to certify that
John Smith
is a
Affiliate
of
The Institution of Fire Engineers
an Association established to promote, encourage and improve the science and practice of Fire Engineering, Fire Prevention and Fire Extinction and all operations and expedients connected therewith, and to give an impulse to ideas likely to be useful in connection with or in relation to such science and practice to the Members of the Institution and to the community at large.

Witness *our hands and Seal at*

Stratford-upon-Avon this 15th *day of* June 2020

International President

Chairman

This certificate is not issued or in pursuance of any statutory authority, or by the authority of the Government or any Department thereof.

Membership Number
000000000

FIGURE 11-5 The Institution of Fire Engineers is a member-driven organization to promote, encourage, and improve the science and practice of fire engineering, fire prevention, and fire extinction.

Wrap-Up

CHAPTER SUMMARY

This chapter has provided an overview of some of the most influential organizations, both private and governmental, that do some type of work in the area of fire research. There are additional organizations and groups that conduct research in the fire industry. There are various paths the researcher can travel when exploring the vast world of fire research. Prevention, mitigation, or suppression activities are areas where the researcher has many opportunities to further the knowledge of fire and fighting fire.

KEY TERMS

Fire Research and Management Exchange System (FRAMES) A systematic method of exchanging information and transferring technology among wildland fire researchers, managers, and other stakeholders.

Fire Research Institute (FRI) A not-for-profit library holding over 80,000 books, journal articles, videos, training manuals, dissertations, news reports, and other materials on wildland fires.

Fire Research Laboratory (FRL) An innovative partnership among law enforcement, fire services, public safety agencies, academia, and the private sector that uses the most advanced scientific, technical, educational, and training methods to make the Bureau of Alcohol, Tobacco, Firearms, and Explosives (ATF) and its partners leaders in fire investigation science to serve and protect the public.

Institution of Fire Engineers An organization with the mission of promoting, encouraging, and improving the science and practice of fire engineering, fire prevention, and fire extinction and all operations and expedients connected therewith and giving impetus to ideas likely to be useful in connection with or in relation to such science and practice to the members of the institution and to the community.

National Fire Data Center An organization involved in the collection, analysis, publication, dissemination, and marketing of information related to the nation's fire problem and U.S. Fire Administration programs.

National Fire Protection Association (NFPA) A member-based organization whose wealth of fire-related research includes investigations of technically significant fire incidents, fire data analysis, and the Charles S. Morgan Technical Library.

National Institute of Standards and Technology (NIST) Develops technologies, measurement methods, and standards.

Underwriters Laboratories (UL) One of the world's leading organizations for product safety testing and certification.

U.S. Fire Administration (USFA) Conducts programs that prevent and mitigate the consequences of fire. The programs are divided into four basic areas: Public Education, Training, Technology, and Data.

REVIEW QUESTIONS

1. You have been assigned the task of conducting research on the number of deaths from fires of children less than 1 year old. Describe the resources that are available to use to conduct this research.
2. Describe the role of NIST as it relates to the fire service.
3. Explain the resources that are available to conduct research on standard fire codes.
4. Identify and explain the resources in your community that have data available to assist you in conducting research.
5. Describe the role of FRAMES.
6. What does the National Fire Data Center do?
7. What type of resources does the NFPA provide?
8. What is NIST?
9. What role does UL fill?
10. What are some of the efforts of the USFA with regard to fire research?

REFERENCES

Almand, K. H. 2008. "The Fire Protection Research Foundation: 25 Years On." http://www.nfpa.org/itemDetail.asp?categoryID=646&itemID=18907&URL=Research%20&%20Reports/Fire%20Protection%20Research%20Foundation/Mission.

ATF Fire Research Laboratory. n.d., http://www.atf.treas.gov/labs/frl.

Fire Research and Management Exchange Systems. n.d., http://frames.nbii.gov/portal/server.pt.

Fire Research Institute. n.d. Accessed June 26, 2020, http://www.fireresearchinstitute.org.

The Institution of Fire Engineers. n.d. Home page. Accessed August 5, 2020, https://www.ife.org.uk/.

National Institute of Standards and Technology. n.d. Fire.gov. Accessed August 5, 2020, https://www.nist.gov/el/fire-research-division-73300/firegov-fire-service.

Underwriters Laboratories. n.d. Accessed June 26, 2020, http://www.ul.com.

U.S. Fire Administration. n.d. Accessed June 26, 2020, http://www.usfa.fema.gov.

© Jones & Bartlett Learning. Photographed by Glen E. Ellman.

Appendix A

National Fallen Firefighters Foundation

Firefighter Life Safety Initiatives

Since the 2004 Summit, the 16 Firefighter Life Safety Initiatives have received broad support throughout the fire service and from related organizations. The 45 member organizations that belong to the National Advisory Committee to the Congressional Fire Services Institute voted unanimously to endorse the 16 initiatives, and several other organizations have taken similar action. The 16 Firefighter Life Safety Initiatives are as follows:

1. Define and advocate the need for a cultural change within the fire service relating to safety; incorporating leadership, management, supervision, accountability and personal responsibility.
2. Enhance the personal and organizational accountability for health and safety throughout the fire service.
3. Focus greater attention on the integration of risk management with incident management at all levels, including strategic, tactical and planning responsibilities.
4. All firefighters must be empowered to stop unsafe practices.
5. Develop and implement national standards for training, qualifications, and certification (including regular recertification) that are equally applicable to all firefighters based on the duties they are expected to perform.
6. Develop and implement national medical and physical fitness standards that are equally applicable to all firefighters, based on the duties they are expected to perform.
7. Create a national research agenda and data collection system that relates to the 16 Firefighter Life Safety Initiatives.
8. Utilize available technology wherever it can produce higher levels of health and safety.
9. Thoroughly investigate all firefighter fatalities, injuries, and near misses.
10. Grant programs should support the implementation of safe practices and procedures and/or mandate safe practices as an eligibility requirement.
11. National standards for emergency response policies and procedures should be developed and championed.
12. National protocols for response to violent incidents should be developed and championed.
13. Firefighters and their families must have access to counseling and psychological support.
14. Public education must receive more resources and be championed as a critical fire and life safety program.
15. Advocacy must be strengthened for the enforcement of codes and the installation of home fire sprinklers.
16. Safety must be a primary consideration in the design of apparatus and equipment. (Everyone Goes Home n.d.)

National Fallen Firefighters Foundation

Congress created the National Fallen Firefighters Foundation (NFFF) to lead a nationwide effort to honor the nation's fallen fire fighters and provide resources to assist their survivors in rebuilding their lives. Since 1992, the nonprofit foundation has developed and expanded programs to fulfill that mandate. The NFFF's new emphasis on preventing line-of-duty deaths is a natural extension of those efforts, which are directed equally toward all fire fighters and involve no other mission or constituency.

When Congress established the NFFF, it provided neither funding nor financial assistance to carry out its mission. However, since 1992, the nonprofit NFFF has developed and expanded programs that fulfill that mandate, as described in the following sections.

Sponsor the Annual National Fallen Firefighters Memorial Weekend

Each October, the NFFF sponsors the official national tribute to all fire fighters who died in the line of duty during the previous year. Thousands attend the weekend activities held at the National Fire Academy in Emmitsburg, Maryland. The weekend features special programs for survivors and coworkers, along with moving public ceremonies.

Help Survivors Attend the Weekend

The NFFF provides travel, lodging, and meals for immediate survivors of fallen fire fighters being honored. This allows survivors to participate in Family Day sessions conducted by trained grief counselors and in the private and public tributes.

Offer Support Programs for Survivors

When a fire fighter dies in the line of duty, the NFFF provides survivors with a place to turn. Families receive emotional assistance through a Fire Service Survivors Network, which matches survivors with similar experiences and circumstances. Families receive a quarterly newsletter and specialized grief publications. The website provides extensive information on survivor benefits, NFFF programs, and other resources.

Award Scholarships to Fire Service Survivors

Spouses, children, and stepchildren of fallen fire fighters are eligible for scholarship assistance for education and job training costs.

Help Departments Deal with Line-of-Duty Deaths

Under a Department of Justice grant, the NFFF offers training to help fire departments handle a line-of-duty death. Departments receive extensive preincident planning support. Immediately after a death, the Chief-to-Chief Network provides technical assistance and personal support to help the department and the family.

Work to Prevent Line-of-Duty Deaths

With the support of fire and life safety organizations, the NFFF has launched a major initiative to reduce fire-fighter deaths. The goal of the Firefighter Life Safety Initiatives Program is to reduce line-of-duty fire-fighter deaths by 25 percent in 5 years and 50 percent within 10 years.

Create a National Memorial Park

The NFFF is expanding the national memorial site in Emmitsburg, Maryland, to create the first permanent national park honoring all fire fighters. The park includes a brick Walk of Honor that connects the Memorial Chapel and the official national monument.

Everyone Goes Home. n.d. "16 Firefighter Life Safety Initiatives." Accessed May 21, 2020, https://www.everyonegoeshome.com/16-initiatives/?/.

National Fallen Firefighters Foundation. n.d. "Home Page." Accessed May 21, 2020, https://www.firehero.org/.

Appendix B

2015 National Fire Service Research Agenda—Recommendations

The final recommendations from the 2015 National Fire Service Research Agenda Symposium include both newly created recommendations and revisions of recommendations from the 2011 National Fire Service Research Agenda Symposium. This appendix lists the 54 final recommendations.

1. Conduct research directed toward identifying those individuals within the fire service who are at a higher risk for specific occupational injury/illness/disease.
2. Continue progress toward the development and refinement of enhanced data systems (such as N-FORS) across all fire service domains.
3. Develop a centralized data warehouse and common data elements to facilitate research related to wildland firefighting.
4. Develop a unified national database with common definitions on fire service fatalities, injuries, and occupational illnesses.
5. Evaluate behavior modification strategies that will lead to lasting cultural changes resulting in improvements in data collection and use.
6. Identify and develop methods to capture operational data on fire-ground performance, mental resiliency, effective communications, and operational benchmarks.
7. Identify and make use of traditional and nontraditional data to supplement, update, and enhance fire service programs, including fire suppression and emergency operations, public education, fire prevention, and community risk reduction efforts.
8. Improve local data collection in order to positively impact efficient service delivery, professional development, and organizational health.
9. Determine the incidence and frequency of occupational diseases/illness/injury/conditions in underrepresented groups and those with unique exposures.
10. Establish a center for best practices for data collection and analysis. Identify and catalog data sources and technology formats that are relevant and beneficial for the fire service.
11. Research total worker health of the wildland firefighter population to improve health and wellness.
12. Create a searchable database of community risk reduction programs that have been identified as best practices in communities and/or states.
13. Assess the effectiveness of risk reduction messages in successfully changing targeted behaviors.
14. Assess the reliability and performance characteristics of alternative smoke alarm technologies.
15. Conduct a cost/benefit analysis of investing in fire department occupational health and safety programs, including identification of best practices and methods to institute such programs.
16. Conduct a study of the life span of PPE.
17. Conduct research based on fire dynamics to identify best practices at the strategic, tactical, and task levels for firefighting operations in new and existing commercial and residential structures. The research should include the creation of on-scene risk assessment tools based on specific fire factors to assist company officers and incident commanders.
18. Conduct research directed toward identifying and overcoming barriers to the implementation of tobacco cessation programs and the elimination of all forms of tobacco and nicotine use (e.g., cigarettes, smokeless tobacco, e-cigarettes, other vape products). Conduct studies related to alcohol abuse, misuse and abuse of prescription drugs and illicit drugs.
19. Conduct research on enhanced dermal protection provided by firefighter structural protective clothing, particularly as it relates to reducing exposures to known and suspected carcinogens.
20. Conduct research on how science can improve wildland firefighting training, tactics, and response to reduce fatalities, injuries, and unintended outcomes.
21. Continue research into operational practices directed toward more effective tactics, improvements in firefighter safety and victim survivability, and reductions in property losses. These studies should specifically address staffing and deployment, fire dynamics research, and victim survivability. The focus should include high-rise residential and commercial buildings, private dwellings, multiple unit residential occupancies, strip malls, taxpayer buildings, and warehouses.
22. Continue research on firefighter health, injury, and diseases related to chronic and repeated exposures to the risks of emergency incidents and the fire service work environment. The research should encompass all disciplines including wildland and wildland–urban interface.
23. Continue research on firefighter health, injury, and diseases related to the risks of acute exposures that may result from emergency incidents.
24. Determine the efficacy/effectiveness of interventions /programs/systems designed to decrease disease/exposure /injury/death and increase medical evaluations, occupational

health, and surveillance. The research should include underresearched populations within the fire service and include a focus on reproductive, maternal, and child health issues; cardiovascular risk factors; injuries; and cancer.

25. Identify respiratory contaminants and determine the potential adverse health outcomes associated with wildland and wildland–urban interface fire operations. Also, determine the adequate respiratory protection for wildland fire fighters.

26. Identify, develop, and refine evidence-based tools and approaches for behavioral health screening, assessment, and intervention.

27. Research the impact of communication failures with portable radio systems and devices as a contributing factor in fire-fighter injuries and fatalities. Address alternative radio system configurations to ensure reliable in-building radio communications. Also, examine potential improvements in radio construction, ergonomics, and the ability to interface portable radios with other technologies to track firefighters in the fire environment.

28. Assess substance misuse and abuse in the fire service (including but not limited to alcohol, prescription drugs, and illicit drugs). Focus research on the identification of effective prevention efforts, interventions, and rehabilitation strategies.

29. Assess the effectiveness of the adoption of codes and standards in reducing the incidence and impact of fires by comparing results in locales that adopted codes and those that did not.

30. Assess the impact of current fire dynamics research on the health of fire investigators.

31. Assess the impact/influence of the adoption/enforcement of codes on the economic impact of wildland–urban interface fires. Consider the impact on both wildland to urban and urban to wildland fire transitions. Examine the impact on both fire ignitions and losses.

32. Assess the overall effectiveness of fire- and injury-reduction programs on the accomplishment of targeted reductions in fatalities, injuries, and property loss.

33. Conduct research on cleaning methods for fire-fighter protective clothing, including potential impacts on the protective properties and useful life of the clothing, and determining effectiveness of removal of suspected carcinogens and other contaminants.

34. Conduct research on the translation, dissemination, and messaging of current knowledge and best practices related to health and wellness programs, including physical fitness, health maintenance, nutrition, and annual medical evaluations.

35. Determine the appropriate level of respiratory protection for use during overhaul operations, including the use of air-monitoring instruments to measure thresholds.

36. Determine the most effective implementation methods to institute occupational health programs.

37. Develop methods to evaluate and quantify the direct and indirect economic impacts of fire service response and operations on property, people, and the environment.

38. Evaluate the impact of modern and evolving building technology (e.g., green buildings, solar and battery storage systems) on fire service operations. Create a knowledge base for incident commanders, company officers, and fire fighters to support operational safely and proficiently.

39. Identify contributing factors to fire-fighter injuries and fatalities related to non–fire-ground events (i.e., EMS, special operations, and roadway).

40. Research the effectiveness of alternative learning mechanisms in order to identify and develop the best firefighter training delivery system(s) for strategic, tactical, and task-level operations.

41. Conduct research into establishing safe and reliable aircraft operations in the wildland–urban interface.

42. Conduct research on the risks and/or benefits of supplements (e.g., nutritional supplements, sports energy drinks, creatine, and testosterone).

43. Conduct research on the efficacy and effectiveness of health and wellness programs for individuals and organizations. Focus on programs directed toward preventive behavioral change. The research areas should include fitness, nutrition, hydration, sleep, and hygiene.

44. Measure the fire growth rate in new homes which are built to modern energy codes and specifications and furnished with contemporary fire loads. Simulate and evaluate escape times based on the realistic capabilities of individuals.

45. Conduct research on the effectiveness of alternative implementation strategies and policies for health and wellness programs.

46. Develop a physical fitness risk assessment tool for wildland fire fighters.

47. Continue to employ fire modeling and full scale re-creations of specific incidents that resulted in firefighter injuries and deaths to identify contributing factors and recommended changes in strategy, tactics, and tasks.

48. Research the development of technology, tactics, and response standards in the wildland–urban interface. Include PPE requirements for all responders.

49. Align research projects with strategies to enhance the benefits of the research. Identify data and technology formats that are most beneficial to the fire service.

50. Conduct research to make improvements in the survivability of fire apparatus crashes. Conduct research related to anthropometric and ergonomic challenges in fire apparatus construction and arrangements that lead to frequent head and musculoskeletal injuries.

51. Develop a user-friendly technological accountability system for use on the fire ground.

52. Evaluate existing ballistic protection options (including helmets) for fire fighters and EMS responders. As indicated by findings, develop new options.

53. Determine the necessary components to be included in the educational process for incident commanders, taking into account risk management, tactics, operational concerns, and an acceptable knowledge base.

54. Research the application of unmanned aerial vehicles for the fire service.

National Fallen Firefighters Foundation. 2016. "Appendix B: 2015 Recommendations." In *2015 National Fire Service Research Agenda: Recommendations Report*. Accessed May 21, 2020, https://www.interagencyboard.org/system/files/resources/2015-Rearch-Agenda-Symposium-Report.pdf.

Appendix C

Fire and Emergency Services Higher Education (FESHE) Correlation Guide

Applications of Fire Research (C0260) Bachelor's (Core) Course Outline	*Applications of Fire Research and Improvement,* Second Edition, Chapter Correlation
Module I: Fundamentals	
1. Consider what research is and why we study it.	1
2. Understand fire-related research objectives.	1
3. Analyze and discuss fire research goals and objectives in relation to the National Institute of Standards and Technology (NIST)-led technical investigation of the World Trade Center disaster.	1
4. Research, evaluate, and discuss sources from which information on fire research is available.	2
5. Identify fire research organizations and programs that have applications to the fire service.	11
6. Identify areas of fire-related research.	1
7. Conduct a preliminary review of current research in a chosen fire-related topic.	2
8. Investigate, evaluate, and interpret research in the area of fire dynamics.	5, 6
9. Investigate, evaluate, and interpret research in the area of fire test standards and codes.	5, 6
Module II: Focusing Your Research Efforts	
1. Define research and its foundations.	2
2. Introduce research methods and approaches.	3, 4
3. Understand the scientific method.	3, 4
4. Conceptualize a strategy for generating research problems.	3, 4
5. Formulate a suitable research problem in an area of fire science.	3, 4

Applications of Fire Research (C0260) Bachelor's (Core) Course Outline	***Applications of Fire Research and Improvement*, Second Edition, Chapter Correlation**
6. Develop a preliminary research proposal outline.	5, 6
7. Distinguish between testing and experimental research.	
8. Compare the results of mathematical fire modeling to full-scale fire testing.	
9. Distinguish between small-, medium-, and large-scale tests and when it is appropriate to use them.	
10. Understand sampling procedures.	
11. Investigate, evaluate, and interpret research in the area of fire safety properties and flammability tests.	
12. Investigate, evaluate, and interpret research in the area of fire modeling.	
Module III: Qualitative Research Methodologies	
1. Develop a familiarity with qualitative research methods and approaches.	8
2. Apply concepts of qualitative methods to fire-related research.	8
3. Select appropriate qualitative methods according to the type of research question raised.	8
4. Interpret conclusions drawn from qualitative methods, based on an analysis of the strengths and weaknesses of the methodology.	8
5. Conduct a literature review related to a fire research problem.	2
6. Investigate, evaluate, and interpret research in the area of structural fire safety.	
7. Investigate, evaluate, and interpret research in the area of life safety.	
8. Investigate, evaluate, and interpret research in the area of firefighter health and safety.	
Module IV: Quantitative Research Methodologies	
1. Develop a familiarity with quantitative research methods and approaches.	6
2. Apply concepts of quantitative methods to fire-related research.	6
3. Apply statistical concepts and data analysis to quantitative research design.	5, 6
4. Select an appropriate quantitative design when the conditions of the research problem demand measurement of variables and relationships.	5, 6

Applications of Fire Research (C0260) Bachelor's (Core) Course Outline	*Applications of Fire Research and Improvement,* Second Edition, Chapter Correlation
5. Select appropriate statistical techniques according to the type of research question raised within a quantitative study.	5, 6
6. Interpret conclusions drawn from statistics as to whether or not they reflect the true properties of phenomena under study.	5, 6
7. Design a research project within a fire research subfield and establish techniques for data gathering and analysis.	5, 6
8. Investigate, evaluate, and interpret research in the area of automatic detection and suppression.	11
9. Investigate, evaluate, and interpret research in the area of transportation fire hazards.	11
10. Investigate, evaluate, and interpret research in the area of risk analysis and loss control.	11
Module V: Applications and Trends in Fire-Related Research	
1. Consider applications of fire-related research to fire safety and prevention.	11
2. Consider future developments in fire-related research.	11
3. Propose specific areas for future research and testing.	11
4. Discuss how your research proposal relates to either applications of fire-related research, future trends in fire-related research, or both.	10
5. Investigate, evaluate, and interpret research in the area of fire service applied research.	10
6. Investigate, evaluate, and interpret research in the area of new trends in fire-related research.	10
7. Complete a formal research proposal in a fire-related field, applying either qualitative or quantitative methods, or a combination of both.	10

Glossary

16 Firefighter Life Safety Initiatives A set of key strategies that must be implemented to meet the U.S. Fire Administration's goal of providing a safer work environment for fire service personnel.

Abstract A very concise and structured summary of a project.

Accreditation phase The phase of organizational improvement in which an independent agency recognizes or affirms that an organization conforms to an established standard.

Alarm When the performance indicator results meet certain conditions, the alarm component of the process control system is triggered.

Analyze phase Determining the causes(s) of problems or what factors may be holding the process back from reaching higher levels of performance.

Association of Public-Safety Communications Professionals (APCO) An organization that sets standards for 911 emergency agencies.

Baldrige National Quality Program A set of criteria used for performance recognition.

Baseline How a process currently operates.

Belmont Report The basis for all federal regulations pertaining to research involving human subjects.

Benchmarking phase The phase of organizational improvement in which an organization begins to transcend the standards it worked so hard to comply with during the regulatory and accreditation phases.

Beneficence The magnitude of potential benefit resulting from a research project or process improvement project should offer some level of justification for the physical, psychological, social, legal, or economic risks it poses to participants.

Bias Effects of extraneous variables on dependent variables.

Blinding Preventing project participants from intentionally biasing the results by preventing them from knowing which version of the process change they are using.

Brainstorming A useful technique for coming up with ideas of potential causes or limiting factors to include in a cause-and-effect diagram.

Capability index A useful tool that quantifies how well the performance of the process met the process standards over a specified period of time.

Case studies Studies that describe a single unit, which may be an event, occurrence, situation, organization, or individual.

Cause-and-effect diagram A tool for identifying potential causes of problems or hindered performance. It is also called a *fishbone* or *Ishikawa diagram*.

Commission on Accreditation for Law Enforcement Agencies (CALEA) An organization that establishes standards for law enforcement agencies.

Commission on Accreditation of Ambulance Services (CAAS) An organization that sets standards for ambulance agencies.

Commission on Fire Accreditation International® (CFAI®) An organization that establishes standards for fire departments.

Compliance officer Someone whose primary responsibility is to ensure compliance with applicable local, state, and federal laws, regulations, and third-party guidelines and who manages audits and investigations into regulatory and compliance issues and responds to requests for information from regulatory bodies.

Control logic The component of the process control system that describes the appropriate actions to take in response to an alarm condition.

Control-limit alarm The presence of a statistical signal that suggests special-cause variation is occurring.

Correlation coefficient Correlation expressed as a value between -1 and 1.

Correlation The strength of a relationship between two variables.

Define phase Clarifies the problem or issue you are trying to address.

Dependent variables The measure of performance (i.e., the performance indicator) that you are trying to change.

DMAIC Define—Measure—Analyze—Improve—Control: A framework used in the Six Sigma methodology.

Ethnography A qualitative research method used for descriptive studies of culture in specific populations or groups that share a common parameter, such as their profession, nationality, religion, location, or even having been exposed to the same event or experience.

Executive summary Covers the highlights and key conclusions of a project.

Extraneous variables Factors, other than the independent variables, that may influence the results seen in the dependent variable.

Fire Research and Management Exchange System (FRAMES) A systematic method of exchanging information and transferring technology among wildland fire researchers, managers, and other stakeholders.

Fire Research Institute (FRI) A not-for-profit library holding over 80,000 books, journal articles, videos, training manuals, dissertations, news reports, and other materials on wildland fires.

Fire Research Laboratory (FRL) An innovative partnership among law enforcement, fire services, public safety agencies, academia, and the private sector that uses the most advanced scientific, technical, educational, and training methods to make the Bureau of Alcohol, Tobacco, Firearms, and Explosives (ATF) and its partners leaders in fire investigation science to serve and protect the public.

Fire Service Performance Indicator Format (FSPIF) model Addresses the issue of standardized formatting for performance indicators and selecting an indicator format.

Full technical report A detailed scientific paper for publication or a detailed internal document.

Grounded-theory study An extension of a phenomenological study.

Histogram Shows the proportion of cases that fall into adjacent, nonoverlapping categories.

Hybrid studies A study that combines ethnography, phenomenology, grounded-theory and constant comparative analysis, and the case study method, and may also include the collection of quantitative data and statistical analysis to characterize data.

Hypothesis The change we make to an independent variable.

Improve phase Uses model programs to evaluate the effectiveness of the program to identify areas to improve.

Improve phrase Choosing the factor to change in an effort to make performance better.

Independent variables Factors that influence process performance.

Institution of Fire Engineers An organization with the mission of promoting, encouraging, and improving the science and practice of fire engineering, fire prevention, and fire extinction and all operations and expedients connected therewith and giving impetus to ideas likely to be useful in connection with or in relation to such science and practice to the members of the institution and to the community.

Institutional review board A committee that reviews applications for research projects, with the goal of protecting the rights and interests of the human subjects who are involved.

Keywords Words used to search for literature pertaining to a certain topic or area of knowledge.

Literature review Used to find previous published research on a topic of interest. It is a means to read about others' findings on a particular subject.

Management dashboard system A library of performance indicators displayed in a format that allows process owners and upper management to see, at a glance, how all of the monitored processes are working, thus allowing managers to have an "early warning system" that will let them know when things are getting off-track.

Measure phase A detailed assessment of the current status of the process (or processes).

Minimum standards Standards with which all professionals or organizations in the industry should be in compliance to be considered competent.

National Academies of Emergency Dispatch (NAED) An organization that sets standards for 911 emergency agencies.

National Fire Data Center An organization involved in the collection, analysis, publication, dissemination, and marketing of information related to the nation's fire problem and U.S. Fire Administration programs.

National Fire Protection Association (NFPA) A member-based organization that sets standards on fire-related topics and whose wealth of fire-related research includes investigations of technically significant fire incidents, fire data analysis, and the Charles S. Morgan Technical Library.

National Institute of Standards and Technology (NIST) Develops technologies, measurement methods, and standards.

P value The means to illustrate the probability that the differences between the groups is the result of chance, not the result of a real difference.

Pareto chart Summarizes and displays graphically the relative importance of the differences between groups of data.

Peer-review process A process that selects those projects that have the most importance and technical merit.

Performance indicator Shows how well (quality) or how efficiently (economically) a process is performing.

Phenomenological research Research that aims to improve our understanding about something that exists in the world we live in.

Placebo A version of the independent variable that has no effect on the dependent variable.

Poster presentations Can include both text and graphics and are usually mounted on large, double-sided poster easels.

Process control plan A way to ensure that favorable improvements made by a process-improvement project will not deteriorate over time.

Process improvement A scientific/data-driven approach to monitoring and improving performance in a particular process.

Qualitative research Developed primarily for use in the social sciences as a method to answer questions about human behavior, culture, attitudes, and the effects of events on people.

Randomization A procedure for randomly choosing which item is used each time the experiment is conducted.

Regression analysis A tool that allows predictions of outcomes to be made on the basis of changes to an input variable.

Regulatory phase The phase of organizational improvement that seeks to ensure that all applicable regulations, laws, contractual requirements, and so forth are being complied with.

Research Scholarly or scientific investigation or inquiry.

Return on investment Projects that have the most important impact for the customer or that have the greatest potential to provide a significant impact on the organization in relation to the money and time invested. ROI may be operational, financial, political, or clinical.

Search engine A website that allows the user to enter keywords to locate documents or other forms of literature on the Internet or in a database or system.

Search string Keywords and phrases that are entered into the search box of a generic or specialized search engine.

Sensor The component of the process control system that serves as a permanent mechanism for measuring process performance, typically consisting of one or more process performance indicators that can monitor process performance over time.

Specification-limit alarm An alarm that is triggered when the performance level falls outside the upper or lower performance specification limit.

Statistical power calculations Calculations tell how often you need to run the test to detect at least a specified level of difference between groups.

Statistically significant As shown by statistical tools, the differences observed between performance before the change and performance after the change (or between one group and another group) that are unlikely to be the result of chance—that is, the differences are real.

U.S. Fire Administration (USFA) Conducts programs that prevent and mitigate the consequences of fire. The programs are divided into four basic areas: Public Education, Training, Technology, and Data.

Underwriters Laboratories (UL) One of the world's leading organizations for product safety testing and certification.

Validation The component of the process control system that documents the new level of process performance after appropriate actions have been taken from the standpoint of process control logic.

Index

Note: Page numbers followed by *f* or *t* indicate material in figures or tables respectively.